Jonathan Lawry

Modelling and Reasoning with Vague Concepts

Studies in Computational Intelligence, Volume 12

Editor-in-chief
Prof. Janusz Kacprzyk
Systems Research Institute
Polish Academy of Sciences
ul. Newelska 6
01-447 Warsaw
Poland
E-mail: kacprzyk@ibspan.waw.pl

Jonathan Lawry

Modelling and Reasoning
with Vague Concepts

 Springer

Dr. Jonathan Lawry
University Bristol
Dept. Engineering Mathematics
University Walk
Queens Building
BRISTOL
UNITED KINGDOM BS8 1TR

Modelling and Reasoning with Vague Concepts

Library of Congress Control Number: 2005935480

ISSN Print Edition: 1860-949X   ISSN Electronic Edition: 1860-9503
ISBN 0-387-29056-7   e-ISBN 0-387-30262-X
ISBN 978-0387-29056-7

Printed on acid-free paper.

Printed in the United States of America.

9 8 7 6 5 4 3 2 1   SPIN 11557296

springeronline.com

For a large class of cases - though not for all - in which we employ the word 'meaning' it can be defined thus: the meaning of a word is its use in language. - Ludwig Wittgenstein

Contents

List of Figures

Preface

Vague concepts are intrinsic to human communication. Somehow it would seems that vagueness is central to the flexibility and robustness of natural lan guage descriptions. If we were to insist on precise concept definitions then we would be able to assert very little with any degree of confidence. In many cases our perceptions simply do not provide sufficient information to allow us to verify that a set of formal conditions are met. Our decision to describe an individual as 'tall' is not generally based on any kind of accurate measurement of their height. Indeed it is part of the power of human concepts that they do not require us to make such fine judgements. They are robust to the imprecision of our perceptions, while still allowing us to convey useful, and sometimes vital, information. The study of vagueness in Artificial Intelligence (AI) is therefore motivated by the desire to incorporate this robustness and flexibility into intelligent computer systems. This goal, however, requires a formal model of vague concepts that will allow us to quantify and manipulate the uncertainty resulting from their use as a means of passing information between autonomous agents.

I first became interested in these issues while working with Jim Baldwin to develop a theory of the probability of fuzzy events based on mass assignments. Fuzzy set theory has been the dominant theory of vagueness in AI since its introduction by Lotfi Zadeh in 1965 and its subsequent successful application in the area of automatic control. Mass assignment theory provides an attractive model of fuzzy sets, but I became increasingly frustrated with a range of technical problems and unintuitive properties that seemed inherent to both theories. For example, it proved to be very difficult to devise a measure of conditional probability for fuzzy sets, that satisfied all of a minimal set of intuitive properties. Also, mass assignment theory provides no real justification for the truth-functionality assumption central to fuzzy set theory.

This volume is the result of my attempts to understand and resolve some of these fundamental issues and problems, in order to provide a coherent framework for modelling and reasoning with vague concepts. It is also an attempt to develop such a framework as can be applied in practical problems concerning automated reasoning, knowledge representation, learning and fusion. I do not believe AI research should be carried out in isolation from potential applications. In essence AI is an applied subject. Instead, I am committed to the idea that theoretical development should be informed by complex practical problems, through the direct application of theories as they are developed. Hence, I have dedicated a significant proportion of this book to presenting the application of the proposed framework in the areas of data analysis, data mining and information fusion, in the hope that this will give the reader at least some indication as to the utility of the more theoretical ideas.

Finally, I believe that much of the controversy in the AI community surrounding fuzzy set theory and its application arises from the lack of a clear operational semantics for fuzzy membership functions, consistent with their truth-functional calculus. Such an interpretation is important for any theory to ensure that its not based on an ad hoc, if internally consistent, set of inference processes. It is also vital in knowledge elicitation, to allow for the translation of uncertainty judgements into quantitative values. For this reason there will be a semantic focus throughout this volume, with the aim of identifying possible operational interpretations for the uncertainty measures discussed.

JONATHAN LAWRY

Acknowledgments

Time is becoming an increasingly rare commodity in this frenetic age. Yet time, time to organise one's thoughts and then to commit them to paper, is exactly what is required for writing a book. For this reason I would like to begin by thanking the Department of Engineering Mathematics at the University of Bristol for allowing me a six month sabbatical to work on this project. Without the freedom from other academic duties I simply would not have been able to complete this volume.

As well as time, any kind of creative endeavour requires a stimulating environment and I would like to thank my colleagues in Bristol for providing just such an environment. I was also very lucky to be able to spend three months during the summer of 2004 visiting the Secció Matemàtiques i Informàtica at the Universidad Politécnica de Cataluña. I would like to thank Jordi Recasens for his kindness during this visit and for many stimulating discussions on the nature of fuzziness and similarity. I am also grateful to the Spanish government for funding my stay at UPC under the scheme 'Ayudas para movilidad de profesores de universidad e investigadores Españoles y extranjeros'.

Over the last few years I have been very fortunate to have had a number of very talented PhD students working on projects relating to the label semantics framework. In particular, I would like to thank Nick Randon and Zengchang Qin who between them have developed and implemented many of the learning algorithms described in the later chapters of this book.

Finally a life with only work would be impoverished indeed and I would like to thank my wonderful family for everything else. To my mother, my wife Pepa, and daughters Ana and Julia - gracias por su amor y su apoyo.

Foreword

Fuzzy set theory, since its inception in 1965, has aroused many controversies, possibly because, for the first time, imprecision, especially linguistic imprecision, was considered as an object of investigation from an engineering point of view. Before this date, there had already been proposals and disputes around the issue of vagueness in philosophical circles, but never before had the vague nature of linguistic information been considered as an important issue in engineering sciences. It is to the merit of Lotfi Zadeh that he pushed this issue to the forefront of information engineering, claiming that imprecise verbal knowledge, suitably formalized, could be relevant in automating control or problem-solving tasks.

Fuzzy sets are simple mathematical tools for modelling linguistic information. Indeed they operate a simple shift from Boolean logic, assuming that there is more to "truth-values" than being true or being false. Intermediate cases, like "half-true" make sense as well, just like a bottle can be half-full. So, a fuzzy set is just a set with blurred boundaries and with a gradual notion of membership. Moreover, the truth-functionality of Boolean logic was kept, yielding a wealth of formal aggregation functions for the representation of conjunction, disjunction and other connectives. This proposal also grounds fuzzy set theory in the tradition of many-valued logics. This approach seems to have generated misunderstandings in view of several critiques faced by the theory of fuzzy sets. A basic reason for the reluctance in established scientific circles to accept fuzzy set theory is probably the fact that while this very abstract theory had an immediate intuitive appeal which prompted the development of many practical applications, the notion of membership functions had not yet been equipped with clear operational semantics. Namely, it is hard to understand the meaning of the number 0.7 on the unit interval, in a statement like "Mr. Smith is tall to degree 0.7", even if it clearly suggests that this person is not tall to the largest extent.

This lack of operational semantics, and of measurement paradigms for membership degrees was compensated for by ad hoc techniques like triangular fuzzy sets, and fuzzy partitions of the reals, that proved instrumental for addressing practical problems. Nevertheless, degrees of membership were confused with degrees of probability, and orthodox probabilists sometimes accused the fuzzy set community of using a mistaken surrogate probability calculus, the main argument being the truth-functionality assumption, which is mathematically inconsistent in probability theory. Besides, there are still very few measurement-theoretic works in fuzzy set theory, while this would be a very natural way of addressing the issue of the meaning of membership grades. Apparently, most measurement-theory specialists did not bother giving it a try.

Important progress in the understanding of membership functions was made by relating fuzzy sets and random sets: while membership functions are not probability distributions, they can be viewed as one-point coverage functions of random sets, and, as such, can be seen as upper probability bounds. This is the right connection, if any, between fuzzy sets and probability. But the price paid is the lack of universal truth-functionality.

The elegant and deep monograph written by Jon Lawry adopts this point of view on membership functions, for the purpose of modelling linguistic scales, with timely applications to data-mining and decision-tree learning. However it adopts a very original point of view. While the traditional random set approach to fuzzy sets considers realisations as subsets of some numerical reference scale (like a scale of heights for "short and tall"), the author assumes they are subsets of the set of labels, obtained from answering yes/no questions about how to label objects. This approach has the merit of not requiring an underlying numerical universe for label semantics. Another highlight of this book is the lucid discussion concerning the truth-functionality assumption, and the proposal of a weaker, yet tractable, "functionality" assumption, where independent atomic labels play a major role. In this framework, many fuzzy connectives can be given an operational meaning. This book offers an unusually coherent and comprehensive, mathematically sound, intuitively plausible, potentially useful, approach to linguistic variables in the scope of knowledge engineering.

Of course, one may object to the author's view of linguistic variables. The proposed framework is certainly just one among many possible other views of membership functions. Especially, one may argue that founding the measurement of gradual entities on yes-no responses to labelling questions may sound like a paradox, and does not properly account for the non-Boolean nature of gradual notions. The underlying issue is whether fuzzy predicates are fuzzy because their crisp extension is partially unknown, or because they are intrinsically gradual in the mind of individuals (so that there just does not exist such a thing as "the unknown crisp extension of a fuzzy predicate"). Although it sounds like splitting hairs, answering this question one way or another has drastic impact on the modelling of connectives and the overall structure of the underlying logic. For instance if "tall" means a certain interval of heights I cannot precisely describe, then "not tall" just means the complement of this interval. So, even though I cannot precisely spot the boundary of the extension of "tall", I can claim that being "tall and not tall" is an outright contradiction, and "being tall or not tall" expresses a tautology. This view enforces the laws of contradiction and excluded-middle, thus forbidding truth-functionality of connectives acting on numerical membership functions. However, if fuzzy predicates are seen as intrinsically gradual, then "tall" and "not tall" are allowed to overlap, then the underlying structure is no longer Boolean and there is room for truth-functionality. Fine, would say the author, but what is the measurement setting that yields such a non-Boolean structure and provides for a clear intuition of membership grades? Such a setting does not exist yet and its discovery remains as an open challenge.

Indeed, while the claim for intrinsically gradual categories is legitimate, most interpretative settings for membership grades proposed so far (random sets, similarity relations, utility ...) seem to be at odds with the truth-functionality assumption, although the latter is perfectly self-consistent from a mathematical point of view (despite what some researchers mistakenly claimed in the past). It is the merit of this book that it addresses the apparent conflict between truth-functionality and operational semantics of fuzzy sets in an upfront way, and that it provides one fully-fledged elegant solution to the debate. No doubt this somewhat provocative but scientifically solid book will prompt useful debates on the nature of fuzziness, and that new alternative proposals will be triggered by its in-depth study. The author must be commended for an extensive work that highlights an important issue in fuzzy set theory, that was perhaps too cautiously neglected by its followers, and too aggressively, sometimes misleadingly, argued about, by its opponents from more established fields.

Didier Dubois,
Directeur de Recherches
IRIT -UPS -CNRS
118 Route de Narbonne
31062 Toulouse Cedex
Toulouse , France

Chapter 1

INTRODUCTION

Every day, in our use of natural language, we make decisions about how to label objects, instances and events, and about what we should assert in order to best describe them. These decisions are based on our partial knowledge of the labelling conventions employed by the population, within a particular linguistic context. Such knowledge is obtained through our experience of the use of language and particularly through the assertions given by others. Since these experiences will differ between individuals and since as humans we are not all identical, our knowledge of labelling conventions and our subsequent use of labels will also be different. However, in order for us to communicate effectively there must also be very significant similarities in our use of label descriptions. Indeed we can perhaps view the labelling conventions of a language as an emergent phenomena resulting from the interaction between similar but subtly different individuals. Now given that knowledge of these conventions is, at best, only partial, resulting as it does from a process of interpolation and extrapolation, we will tend to be uncertain about how to label any particular instance. Hence, labelling decisions must then be made in the presence of this uncertainty and based on our limited knowledge of language rules and conventions. A consequence of this uncertainty is that individuals will find it difficult to identify the boundaries amongst instances at which concepts cease to be applicable as valid descriptions.

The model of vague concepts presented in this volume is fundamentally linked to the above view of labelling and the uncertainty associated with the decisions that an intelligent agent must make when describing an instance. Central to this view is the assumption that agents believe in the meaningfulness of these decisions. In other words, they believe that there is a 'correct way' to use words in order to convey information to other agents who share the same (or similar) linguistic background. By way of justification we would argue that such a stance is natural on the part of an agent who must make crisp decisions about what words and labels to use across a range of contexts. This view would seem to be consistent with the epistemic model of vagueness proposed by Williamson [108] but where the uncertainty about concept definition is identified as being linguistic in nature. On the other hand, there does seem to be a subtle difference between the two positions in that our view does not assume the actual existence of some objectively correct (but unknown) definition of a vague concept. Rather individuals assume that there is a fixed (but partially known) labelling convention to which they should adhere if they wish to be understood. The actual rules for labelling then emerge from the interaction between individuals making such an assumption. Perhaps we might say that agents find it useful to adopt an 'epistemic stance' [1] regarding the applicability of vague concepts. In fact our view seems closer to that of Parikh [77] when he argues for an 'anti-representational' view of vagueness based on the use of words through language games.

We must now pause to clarify that this volume is not intended to be primarily concerned with the philosophy of vagueness. Instead it is an attempt to develop a formal quantitative framework to capture the uncertainty associated with the use of vague concepts, and which can then be applied in artificial intelligence systems. However, the underlying philosophy is crucial since in concurrence with Walley [102] we believe that, to be useful, any formal model of uncertainty must have a clear operational semantics. In the current context this means that our model should be based on a clearly defined interpretation of vague concepts. A formal framework of this kind can then allow for the representation of high-level linguistic information in a range of application areas. In this volume, however, we shall attempt to demonstrate the utility of our framework by focussing particularly on the problem of learning from data and from background knowledge.

In many emerging information technologies there is a clear need for automated learning from data, usually collected in the form of large databases. In an age where technology allows the storage of large amounts of data, it is necessary to find a means of extracting the information contained to provide useful models. In machine learning the fundamental goal is to infer robust models with good generalization and predictive accuracy. Certainly for some applications this is all that is required. However, it is also often required that the learnt models should be relatively easy to interpret. One should be able to understand the rules or procedures applied to arrive at a certain prediction or decision. This is particularly important in critical applications where the consequences of a wrong decision are extremely negative. Furthermore, it may be that some kind of qualitative understanding of the system is required rather than simply a 'black box' model that can predict (however accurately) it's behaviour. For example, large companies such as supermarkets, high street stores and banks continuously collect a stream of data relating to the behaviour of their customers. Such data must be analysed in such a way as to give an understanding of important trends and relationships and to provide flexible models that can be used for a range of decision making tasks. From this perspective a representational framework based on first order logic combined with a model of the uncertainty associated with using natural language labels, can provide a very useful tool. The high-level logical language means that models can be expressed in terms of rules relating different parameters and attributes. Also, the underlying vagueness of the label expressions used allows for more natural descriptions of the systems, for more robust models and for improved generalisation.

In many modelling problems there is significant background knowledge available from domain experts. If this can be elicited in an appropriate form and then fused with knowledge inferred from data then this can lead to significant improvements in the accuracy of the models obtained. For example, taking into account background knowledge regarding attribute dependencies, can often simplify the learning process and allow the use of simpler, more transparent, models. However, the process of knowledge elicitation is notoriously difficult, particularly if knowledge must be translated into a form unfamiliar to the expert. Alternatively, if the expert is permitted to provide their information as rules of thumb expressed in natural language then this gives added flexibility in the elicitation process. By translating such knowledge into a formal framework we can then investigate problems of inductive reasoning and fusion in a much more conceptually precise way.

The increased use of natural language formalisms in computing and scientific modelling is the central goal of Zadeh's 'computing with words' programme [117]. Zadeh proposes the use of an extended constraint language, referred to as precisiated natural language [118], and based fundamentally on fuzzy sets. Fuzzy set theory and fuzzy logic, first introduced by Zadeh in [110], have been the dominant methodology for modelling vagueness in AI for the past four or five decades, and for which there is now a significant body of research literature investigating both formal properties and a wide range of applications. Zadeh's framework introduces the notion of a linguistic variable [112]-[114], defined to be a variable that takes as values natural language terms such as *large, small, medium* etc and where the meaning of these words is given by fuzzy sets on some underlying domain of discourse. An alternative linguistic framework has been proposed by Schwartz in a series of papers including [92] and [93]. This methodology differs from that of Zadeh in that it is based largely on inference rules at the symbolic level rather than on underlying fuzzy sets. While the motivation for the framework proposed in this volume is similar to the computing with words paradigm and the work of Schwartz, the underlying calculus and its interpretation are quite different. Nonetheless, given the importance and success of fuzzy set theory, we shall throughout make comparisons between it and our new framework.

In chapter 2 we overview the use of fuzzy set theory as a framework for describing vague concepts. While providing a brief introduction to the basic mathematics underlying the theory, the main focus of the chapter will be on the interpretation or operational semantics of fuzzy sets rather than on their formal properties. This emphasis on semantics is motivated by the conviction that in order to provide an effective model of vagueness or uncertainty, the measures associated with such a framework must have a clearly understood meaning. Furthermore, this meaningful should be operational, especially in the sense that it aids the elicitation of knowledge and allows for the expression of clearly interpretable models. In particular, we shall discuss the main interpretations of fuzzy membership functions that have been proposed in the literature and consider whether each is consistent with a truth-functional calculus like that proposed by Zadeh [110]. Also, in the light of results by Dubois and Prade [20], we shall emphasise the strength of the truth-functionality assumption itself and suggest that a weaker form of functionality may be more appropriate. Overall we shall take the view that the concept of fuzzy sets itself has a number of plausible interpretations but none of these provides an acceptable justification for the assumption of truth-functionality.

Chapter 3 introduces the label semantics framework for modelling vague concepts in AI. This attempts to formalise many of the ideas outlined in this first part of this chapter by focusing on quantifying the uncertainty that an intelligent agent has about the labelling conventions of the population in which he/she is a member, and specifically the uncertainty about what labels are appropriate to describe any particular given instance. This is achieved through the introduction of two strongly related measures of uncertainty, the first quantifying the agents belief that a particular expression is appropriate to describe an instance and the second quantifying the agents uncertainty about which amongst the set of basic labels, are appropriate to describe the instance. Through the interaction between these two measures it is shown that label semantics can provide a functional but never truth-functional calculi, with which to reason about vague concept labels. The functionality of such a calculus can be related to some of the combination operators for conjunction and disjunction used in fuzzy logic, however, in label semantics such operators can only be applied to simple conjunctions and disjunctions of labels and not to more complex logical expressions. Also, in chapter 3 it is shown how this framework can be used to investigate models of assertion, whereby an agent must choose what particular logical expression to use in order to describe an instance. This must take account of the specificity and logical form of the expression as well as its level of appropriateness as a description. Finally, in this chapter we will relate the label semantics view of vagueness to a number of other theories of vagueness proposed within the philosophy literature.

The version of label semantics outlined in chapter 3 is effectively 1-dimensional in that it is based on the assumption that each object is described in terms of only one attribute. In chapter 4 we extend the theory to allow for multiple attributes and also by allowing descriptions of more than one object. Also in this chapter we show how multi-dimensional label expressions (object descriptions referring to more than one attribute) can be used to model input-output relationships in complex systems. Finally, by extending our measures of uncertainty to descriptions of multiple objects we consider how such measures can be aggregated across a database. This will have clear applications in chapters 6 and 7 where we apply our framework to data modelling and learning.

Another fundamental consideration regarding the use of vague concepts concerns the information that they provide about the object or objects being described. If we are told that 'Bill is *tall*' what information does this give us about Bill and specifically about Bill's height? In chapter 5 we investigate a number of different theories of the information provided by vague concepts. A number of these relate to fuzzy set theory and include Zadeh's possibility theory [115] and probability theory of fuzzy events [111]. Chapter 5 also presents a number of alternative models of probability conditional on vague information, based on both mass assignment theory [5]-[7] and on the label semantics framework introduced in chapters 3 and 4. In addition to the information provided about objects we also consider the information provided by vague concepts about other vague concepts. In this context we investigate measures of conditional matching of expressions and in particular conditional probability. For the latter a number of conditional probability measures are proposed based on Zadeh's framework, mass assignment theory and label semantics. These are then assessed in terms of whether or not they satisfy a number of axioms and properties characterising certain intuitive epistemic principles. Finally, in the label semantics model we consider the case of conditioning on knowledge taking the form of a distribution (mass assignment) across sets of possible labels.

Chapters 6 and 7 concern the use label semantics as a representation framework for data modelling. Chapter 6 focuses on the development of learning algorithms for both classification and prediction (sometimes called regression) tasks. Both types of problem require the learning of input-output functional relationships from data, where for classification problems the outputs are discrete classes and for prediction problems the outputs are real numbers. Two types of linguistic models are introduced, capable of carrying out both types of task. Mass relations quantify the link between sets of appropriate labels on inputs and either output classes or appropriate label sets for describing a real valued output. Linguistic decision trees are tree structured sets of quantified rules, where for the each rule the antecedent is a conjunction of label expressions describing input attributes and the consequent is either a class or an expression describing the real-valued output. Associated with each such rule is the conditional probability of the consequent given the antecedent. Consequently rules of this form can model both concept vagueness and uncertainty inherent in the training data.

Chapter 7 investigates the use of label semantics framework for data and knowledge fusion. For example, it is shown how the conditional measures introduced in chapter 5 can be used to generate informative prior distributions on the domain of discourse from background knowledge expressed in terms of linguistic statements. This approach can be extended to the case where the available knowledge is uncertain. In such cases a family of priors are identified and additional constraints must be introduced in order to obtain a unique solution. In the second part of chapter 7 we focus on the use of mass relations to fuse background knowledge and data. This again incorporates conditioning methods described in chapter 5. In this context we investigate fusion in classification problems and also in reliability analysis for engineering systems. The latter is particularly relevant since although empirical data is usually limited there is often other qualitative information available. Overall we demonstrate that the incorporation of appropriate background knowledge results in more accurate and informative models.

In Chapter 8 we return to consider theoretical issues concerning the label semantics calculus, by introducing non-additive measures of appropriateness. These provides a general theory incorporating a wider range of measures and allowing for a more flexible, if somewhat more abstract, calculus. This chapter does show that the assumption of additivity is not unavoidable in the development of the label semantics framework. However, the application of these generalised measures in AI remains to be investigated.

Notes

1 This terminology is inspired by Dennett's idea of an 'intentional stance' in [18].

Chapter 2

VAGUE CONCEPTS AND FUZZY SETS

Vague or fuzzy concepts are fundamental to natural language, playing a central role in communications between individuals within a shared linguistic context. In fact Russell [90] even goes so far as to claim that all natural language concepts are vague. Yet often vague concepts are either viewed as problematic because of their susceptibility to Sorites paradoxes or at least as somehow 'second rate' when compared with the more precise concepts of the physical sciences, mathematics and formal logic. This view, however, does not properly take account of the fact that vague concepts seem to be an effective means of communicating information and meaning. Sometimes more effective, in fact, than precise alternatives. Somehow, knowing that 'the robber was *tall*' is more useful to the police patrolling the streets, searching for suspects, than the more precise knowledge that 'the robber was *exactly 1.8 metres* in height'. But what is the nature of the information conveyed by fuzzy statements such as 'the robber was *tall*' and what makes it so useful? It is an attempt to answer this and other related questions that will be the central theme of this volume. Throughout, we shall unashamedly adopt an Artificial Intelligence perspective on vague concepts and not even attempt to resolve longstanding philosophical problems such as Sorites paradoxes. Instead, we will focus on developing an understanding of how an intelligent agent can use vague concepts to convey information and meaning as part of a general strategy for practical reasoning and decision making. Such an agent could be an artificial intelligence program or a human, but the implicit assumption is that their use of vague concepts is governed by some underlying internally consistent strategy or algorithm. For simplicity this agent will be referred to using the pronoun You. This convention is borrowed from Smets work on the Transferable Belief Model (see for example [97]) although the focus of this work is quite different. We shall immediately attempt to reduce the enormity of our task by restricting the type of vague concept to be considered. For the purposes of this volume we shall restrict our attention to concepts as identified by words such as adjectives or nouns that can be used to describe a object or instance. For such an expression θ it should be meaningful to state that 'x is θ' or that 'x is a θ'[1]. Given a universe of discourse Ω containing a set of objects or instances to be described, it is assumed that all relevant expression can be generated recursively from a finite set of basic labels LA. Operators for combining expressions are restricted to the standard logical connectives of negation ($\neg$), conjunction ($\wedge$), disjunction ($\vee$) and implication ($\rightarrow$). Hence, the set of label expressions identifying vague concepts can be formally defined as follows:

DEFINITION 1 *Label Expressions*
The set of label expressions of LA, LE, is defined recursively as follows:

(i) $L \in LE, \forall L \in LA$

(ii) If $\theta, \varphi \in LE$ then $\neg\theta, \theta \wedge \varphi, \theta \vee \varphi, \theta \rightarrow \varphi \in LE$

For example, Ω could be the set of suspects for a robbery and LA might correspond to a set of basic labels used by police for identifying individuals, such as $LA =$ $\{tall, \ medium, \ short, \ medium \ build, \ heavy \ build, \ stocky, \ thin, \ldots$ $blue \ eyes, \ brown \ eyes \ldots\}$. In this case possible expressions in LE include $medium \wedge \neg tall \wedge brown \ eyes$ ('medium but not tall with brown eyes') and $short \wedge (medium \ build \vee heavy \ build)$ ('short with medium or heavy build').

Since it was first proposed by Zadeh in 1965 [110] the treatment of vague concepts in artificial intelligence has been dominated by fuzzy set theory. In this volume, we will argue that aspects of this theory are difficult to justify, and propose an alternative perspective on vague concepts. This in turn will lead us to develop a new mathematical framework for modelling and reasoning with imprecise concepts. We begin, however, in this first chapter by reviewing current theories of vague concepts based on fuzzy set theory. This review will take a semantic, rather than purely axiomatic, perspective and investigate a number of proposed operational interpretations of fuzzy sets, taking into account their consistency with the mathematical calculus of fuzzy theory.

2.1   Fuzzy Set Theory

The theory of fuzzy sets, based on a truth-functional calculus proposed by Zadeh [110], is centred around the extension of classical set theoretic operations such as union and intersection to the non-binary case. Fuzzy sets are generalisations of classical (crisp) sets that allow elements to have partial membership. Every crisp set A is characterised by its membership function $\chi_A : \Omega \rightarrow \{0, 1\}$ where $\chi_A(x) = 1$ if and only if $x \in A$ and where $\chi_A(x) = 0$ otherwise. For fuzzy sets this definition is extended so that $\chi_A : \Omega \to [0, 1]$ allowing x to have partial membership $\chi_A(x)$ in A.

Fuzzy sets can be applied directly to model vague concepts through the notion of extension. The extension of a crisp (non-fuzzy) concept θ is taken to be those objects in the universe Ω which satisfy θ i.e $\{x \in \Omega : \text{'}x \text{ is } \theta\text{'} \text{ is true}\}$. In the case of vague concepts it is simply assumed that some elements have only partial membership in the extension. In other words, the extension of a vague concept is taken to be a fuzzy set. Now in order to avoid any cumbersome notation we shall also use θ to denote the extension of an expression $\theta \in LE$. Hence, according to fuzzy set theory [110] the extension of a vague concept θ is defined by a fuzzy set membership function $\chi_\theta : \Omega \to [0, 1]$. Now given this possible framework You are immediately faced with a difficult computational problem. Even for a finite basic label set LA there are infinitely many expressions in LE generated by the recursive definition 1. You cannot hope to explicitly define a membership function for any but a small subset of these expressions. Fuzzy set theory attempts to overcome this problem by providing a mechanism according to which the value for $\chi_\theta(x)$ can be determined uniquely from the values $\chi_L(x) : L \in LE$. This is achieved by defining a mapping function for each of the standard logical connectives; $f_\wedge : [0, 1]^2 \to [0, 1]$, $f_\vee : [0, 1]^2 \to [0, 1]$, $f_\to : [0, 1]^2 \to [0, 1]$ and $f_\neg : [0, 1] \to [0, 1]$. The value of $\chi_\theta(x)$ for any expression θ and value $x \in \Omega$ can then be determined from $\chi_L(x) : L \in LA$ according to the following recursive rules:

$$\forall \theta, \varphi \in LE \; \chi_{\theta \wedge \varphi}(x) = f_\wedge(\chi_\theta(x), \chi_\varphi(x))$$

$$\forall \theta, \varphi \in LE \; \chi_{\theta \vee \varphi}(x) = f_\vee(\chi_\theta(x), \chi_\varphi(x))$$

$$\forall \theta, \varphi \in LE \; \chi_{\theta \to \varphi}(x) = f_\to(\chi_\theta(x), \chi_\varphi(x))$$

$$\forall \theta \in LE \; \chi_{\neg\theta}(x) = f_\neg(\chi_\theta(x))$$

This assumption is referred to as truth-functionality [2] due to the fact that it extends the recursive mechanism for determining the truth-values of compound sentences from propositional variables in propositional logic. In fact, a fundamental assumption of fuzzy set theory is that the above functions coincide with the classical logic operators in the limit case when $\chi_\theta(x)$, $\chi_\varphi(x) \in \{0, 1\}$. Beyond this constraint it is somewhat unclear as to what should be the precise definition of these combination functions. However, there is a wide consensus that $f_\wedge$, $f_\vee$ and $f_\neg$ [54] should satisfy the following sets of axioms:

Conjunction

C1 $\forall a \in [0, 1] \; f_\wedge(a, 1) = a$

C2 $\forall a, b, c \in [0, 1]$ if $b \leq c$ then $f_\wedge(a, b) \leq f_\wedge(a, c)$

C3 $\forall a, b \in [0, 1] \; f_\wedge (a, b) = f_\wedge (b, a)$

C4 $\forall a, b, c \in [0, 1] \; f_\wedge (f_\wedge (a, b), c) = f_\wedge (a, f_\wedge (b, c))$

Disjunction

D1 $\forall a \in [0, 1] \; f_\vee (a, 0) = a$

D2 $\forall a, b, c \in [0, 1] \;$ if $b \leq c$ then $f_\vee (a, b) \leq f_\vee (a, c)$

D3 $\forall a, b \in [0, 1] \; f_\vee (a, b) = f_\vee (b, a)$

D4 $\forall a, b, c \in [0, 1] \; f_\vee (f_\vee (a, b), c) = f_\vee (a, f_\vee (b, c))$

Negation

N1 $f_\neg (1) = 0$ and $f_\neg (0) = 1$

N2 $f_\neg$ is a continuous function on $[0, 1]$

N3 $f_\neg$ is a decreasing function on $[0, 1]$

N4 $\forall a \in [0, 1] \; f_\neg (f_\neg (a)) = a$

Axioms C1-C4 mean that $f_\wedge$ is a triangular norm or (t-norm) as defined by [94] in the context of probabilistic metric spaces. Similarly according to D1-D4 $f_\vee$ is a triangular conorm (t-conorm). An infinite family of functions satisfy the t-norm and t-conorm axioms including $f_\wedge = \min$ and $f_\vee = \max$ proposed by Zadeh [110]. Other possibilities are, for conjunction, $f_\wedge (a, b) = a \times b$ and $f_\wedge (a, b) = \max (0, a + b - 1)$ and, for disjunction, $f_\vee (a, b) = a + b - a \times b$ and $\min (1, a + b)$. Indeed it can be shown [54] that $f_\wedge$ and $f_\vee$ are bounded as follows:

$$\forall a, b \in [0, 1] \; \underline{f_\wedge} (a, b) \leq f_\wedge (a, b) \leq \min (a, b)$$
$$\forall a, b \in [0, 1] \; \max (a, b) \leq f_\vee (a, b) \leq \overline{f_\vee} (a, b)$$

where $\underline{f_\wedge}$ is the drastic t-norm defined by:

$$\forall a, b \in [0, 1] \; \underline{f_\wedge} (a, b) = \begin{cases} a & : & b = 1 \\ b & : & a = 1 \\ 0 & : & \text{otherwise} \end{cases}$$

and $\overline{f_\vee}$ is the drastic t-conorm defined by:

$$\forall a, b \in [0, 1] \; \overline{f_\vee} (a, b) = \begin{cases} a & : & b = 0 \\ b & : & a = 0 \\ 1 & : & \text{otherwise} \end{cases}$$

Interestingly, adding the additional idempotence axioms restricts t-norms to min and t-conorms to max:

Idempotence Axioms

C5 $\forall a \in [0,1] \; f_\wedge (a,a) = a$

D5 $\forall a \in [0,1] \; f_\vee (a,a) = a$

THEOREM 2 $f_\wedge$ *satisfies C1-C5 if and only if* $f_\wedge = \min$
Proof
($\Leftarrow$) Trivially min *satisfies C1-C5*
($\Rightarrow$) Assume $f_\wedge$ *satisfies C1-C5*
For $a, b \in [0,1]$ *suppose* $a \le b$ *then*

$$a = f_\wedge (a,a) \le f_\wedge (a,b) \le f_\wedge (a,1) = a$$

by axioms C1, C2 and C5 and therefore $f_\wedge (a,b) = a = \min(a,b)$
Alternatively, for $a, b \in [0,1]$ *suppose* $b \le a$ *then*

$$b = f_\wedge (b,b) \le f_\wedge (a,b) \le f_\wedge (1,b) = b$$

by axioms C1, C2, C3 and C5 and therefore $f_\wedge (a,b) = b = \min(a,b)$ $\square$

THEOREM 3 $f_\vee$ *satisfies D1-D5 if and only if* $f_\vee = \max$

The most common negation function $f_\neg$ proposed is $f_\neg (a) = 1 - a$ although there are again infinitely many possibilities including, for example, the family of parameterised functions defined by:

$$f_\neg^\lambda (a) = \frac{1 - a}{1 + \lambda a}$$

Somewhat surprisingly, however, all negation functions essentially turn out to be rescalings of $f_\neg (a) = 1 - a$ as can be seen from the following theorem due to Trillas [101].

THEOREM 4   *If* $f_\neg$ *satisfies N1-N4 then* $\langle [0,1], f_\neg, < \rangle$ *is isomorphic to* $\langle [0,1], 1 - x, < \rangle$
Proof
Since $f_\neg (1) = 0$ *and* $f_\neg (0) = 1$ *then by continuity (N2) it follows that there is some value* k *such that* $f_\neg (k) = k$ *(see figure 2.1). Now define*

$$g(x) = \begin{cases} \frac{x}{2k} & : \quad x \le k \\ 1 - \frac{f_\neg(x)}{2k} & : \quad x > k \end{cases}$$

Then $g(0) = 0$, $g(1) = 1$ and $g(k) = 0.5$. Also, it is easy to check that g is strictly increasing and continuous and hence onto.

Finally, for $x \leq k$ $f_\neg(x) \geq f_\neg(k) = k$ (by N3) and therefore

$$g(f_\neg(x)) = 1 - \frac{f_\neg(f_\neg(x))}{2k} = 1 - \frac{x}{2k} = 1 - g(x) \ by \ N4$$

Similarly, for $x > k$ then $f_\neg(x) \leq f_\neg(k) = k$ and therefore

$$g(f_\neg(x)) = \frac{f_\neg(x)}{2k} = 1 - g(x) \ as \ required. \ \square$$

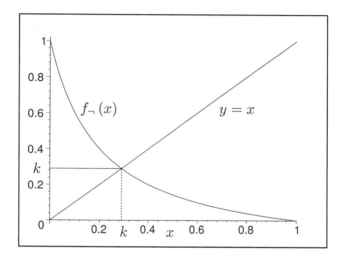

Figure 2.1: Plot of a possible $f_\neg$ function and its associated k value

In view of this rather strong result we shall now assume that $f_\neg(a) = 1 - a$ and move on to consider possible relationships between t-norms and t-conorms. Most of the constraints relating t-norms and t-conorms come from the imposition of classical logic equivalences on vague concepts. Typical of these is the duality relationship that emerges from the assumption that vague concepts satisfy de Morgan's Law i.e. that $\theta \vee \varphi \equiv \neg\theta \wedge \neg\varphi$. In the context of truth-functional fuzzy set theory this means that:

$$\forall a, b \in [0, 1] \ f_\vee(a, b) = 1 - f_\wedge(1 - a, 1 - b)$$

Accordingly the table in figure 2.2 shows a number of well known t-norms with their associated t-conorm duals:

Another obvious choice of logical equivalence that we might wish vague concepts to preserve is $\theta \wedge \theta \equiv \theta$ (similarly $\theta \vee \theta \equiv \theta$). In terms of t-norms (t-conorms) this leads to the idempotence axiom C5 (D5) which as we have seen from theorem 2 (theorem 3) restricts us to min (max).

t-norm $f_\wedge (a, b)$	t-conorm dual $f_\vee (a, b)$
$\min (a, b)$	$\max (a, b)$
$a \times b$	$a + b - a \times b$
$\max (0, a + b - 1)$	$\min (1, a + b)$
$\underline{f_\wedge (a, b)}$	$\overline{f_\vee (a, b)}$

Figure 2.2: t-norms with associated dual t-conorms

In additional, to constraints based on classical logical equivalences it might also be desirable for fuzzy memberships to be additive in the sense that

$$\forall x \in \Omega \; \forall \theta, \varphi \in LE \; \chi_{\theta \vee \varphi} (x) = \chi_\theta (x) + \chi_\varphi (x) - \chi_{\theta \wedge \varphi} (x)$$

This generates the following equation relating t-norms and t-conorms:

$$\forall a, b \in [0, 1] \; f_\vee (a, b) = a + b - f_\wedge (a, b)$$

Making the additional assumption that $f_\vee$ is the dual of $f_\wedge$ we obtain Frank's equation [31]:

$$\forall a, b \in [0, 1] \; f_\wedge (a, b) - f_\wedge (1 - a, 1 - b) = a + b + 1$$

Frank [31] showed that for $f_\wedge$ to satisfy this equation it must correspond to an ordinal sum of members of the following parameterised family of t-norms:

DEFINITION 5 *For parameter* $s \in [0, \infty)$

$$f_{\wedge,s}(y_1, y_2) = \begin{cases} \min(y_1, y_2) & : \quad s = 0 \\ \log_s \left(1 + \frac{(s^{y_1} - 1)(s^{y_2} - 1)}{s - 1}\right) & : \quad s > 0, \; s \neq 1 \\ y_1 \times y_2 & : \quad s = 1 \end{cases}$$

In Frank's t-norms the parameter s in $f_{\wedge,s} (\chi_\theta(x), \chi_\varphi(x))$ effectively provides a measure of the dependence between the membership functions of the expressions θ and φ. The smaller the value of s the stronger the dependence (see for example [2] or [46]).

2.2 Functionality and Truth-Functionality

As described in the previous section fuzzy logic [110] is truth-functional, a property which significantly reduces both the complexity and storage requirements of the calculus. Truth-functionality is, however, a rather strong assumption that significantly reduces the number of standard Boolean properties that can be satisfied by the calculus. For instance, Dubois and Prade

[20] effectively showed that no non-trivial truth-functional calculus can satisfy idempotence together with the law of excluded middle. ,

THEOREM 6 *Dubois and Prade [20]*
If χ is truth-functional and satisfies both idempotence and the law of excluded middle then $\forall \theta \in LE \; \forall x \in \Omega \; \chi_\theta(x) \in \{0, 1\}$
Proof
From theorem 3 we have that the only idempotent t-conorm is max. *Now $\forall \theta \in LE \; \forall x \in \Omega \; \chi_{\neg\theta}(x) = f_\neg(\chi_\theta(x))$. Hence, by the law of excluded middle* max $(\chi_\theta(x), f_\neg(\chi_\theta(x))) = 1$. *Now if $\chi_\theta(x) = 1$ then the result is proven. Otherwise if $f_\neg(\chi_\theta(x)) = 1$ then by negation axiom N4 $\chi_\theta(x) = f_\neg(f_\neg(\chi_\theta(x))) = f_\neg(1) = 0$ by negation axiom N1 as required.*

Elkan [30] somewhat controversially proved a related result for Zadeh's original min / max calculus. Elkan's result focuses on the restrictions imposed on membership values if this calculus is to satisfy a particular classical equivalence relating to re-expressions of logical implication. Clearly this theorem is weaker than that of Dubois and Prade [20] in that it only concerns one particular choice of t-norm, t-conorm and negation function.

THEOREM 7 *Elkan [30]*
Let $f_\wedge(a, b) = \min(a, b)$, $f_\vee(a, b) = \max(a, b)$ and $f_\neg(a) = 1 - a$. For this calculus if $\forall \theta, \varphi \in LE$ and $\forall x \in \Omega \; \chi_{\neg(\theta \wedge \neg\varphi)}(x) = \chi_{\varphi \vee (\neg\theta \wedge \neg\varphi)}(x)$ then $\forall \theta, \varphi \in LE$ and $\forall x \in \Omega$ either $\chi_\theta(x) = \chi_\varphi(x)$ or $\chi_\varphi(x) = 1 - \chi_\theta(x)$

The controversy associated with this theorem stems mainly from Elkan's assertion in [30] that such a result means that previous successful practical applications of fuzzy logic are somehow paradoxical. The problem with Elkan's attack on fuzzy logic is that it assumes a priori that vague concepts should satisfy a specific standard logical equivalence, namely $\neg (\theta \wedge \neg\varphi) \equiv \varphi \vee (\neg\theta \wedge \neg\varphi)$. No justification is given for the preservation of this law in the case of vague concepts, except that it corresponds to a particular representation of logical implication. Given such an equivalence and assuming the truth-functional min / max calculus of Zadeh then the above reduction theorem (theorem 7) follows trivially.

In their reply to Elkan, Dubois etal. [23] claim that he has confused the notions of epistemic uncertainty and degree of truth. Measures of epistemic uncertainty, they concede, should satisfy the standard Boolean equivalences while degrees of truth need not. In one sense we agree with this point in that Elkan seems to be confusing modelling the uncertainty associated with the object domain Ω with modelling the vagueness of the concepts in the underlying description language LE. On the other hand we do not agree that you can completely separate these two domains. When You make assertions involving vague concepts then Your intention is to convey information about Ω (a fact recognized in fuzzy set theory by the linking of fuzzy membership functions with possibility distributions [115]). It does not seem reasonable that questions related to this process should be isolated from those relating to the underlying calculus for combining vague concepts. The way in which You conjunctively combine two concepts L_1 and L_2 must be dependent on the information You want to convey about x when You assert 'x is $L_1 \wedge L_2$' and the relationship of this information with that conveyed by the two separate assertions 'x is L_1' and 'x is L_2'. Furthermore, it is not enough to merely state that truth-degrees are different from uncertainty and are (or can be) truth-functional. Rather, we claim that the correct approach is to develop an operational semantics for vague concepts and investigate what calculi emerge. Indeed this emphasis on a semantic approach forms the basis of our main object to Elkan's work. The problem of what equivalences must be satisfied by vague concepts should be investigated within the context of a particular semantics. It is not helpful to merely select such an equivalence largely arbitrarily and then proceed as if the issue had been resolved.

The theme of operation semantics for vague concepts is one that we will return to in a later section of this chapter and throughout this volume. However, for the moment we shall take a different perspective on the result of Dubois and Prade (and to a lesser degree that of Elkan) by noting that it provides an insight into what a strong assumption truth-functionality actually is. We also suggest that truth-functionality is a special case of a somewhat weaker assumption formalizing the following property: Functionality [3] assumes that for any sentence $\theta \in LE$ there exists some mechanism by which $\forall x \in \Omega$ $\chi_\theta(x)$ can be determined only from the values $\{\chi_L(x) : L \in LA\}$ (i.e. the membership values of x across the basic labels). This notion seems to capture the underlying intuition that the meaning of compound vague concepts are derived only from the meaning of their component concepts while, as we shall see in the sequel, avoiding the problems highlighted by the theorems of Dubois and Prade and of Elkan.

DEFINITION 8 *The measure ν on $LE \times \Omega$ is said to be functional if $\forall \theta \in LE$ there is function $f_\theta : [0,1]^n \to [0,1]$ such that $\forall x \in \Omega$ $\nu_\theta(x) = f_\theta(\nu_{L_1}(x), \ldots, \nu_{L_n}(x))$*

The following example shows that functional measures are not necessarily subject to the triviality result of Dubois and Prade [20].

EXAMPLE 9 *Functional but Non-Truth Functional Calculus*
Let $LA = \{L_1, L_2\}$ and for $\theta \in LE$ let θ_x denote the proposition 'x is θ'. Now let $\nu_\theta(x)$ denote $P(\theta_x)$ where P is a probability measure on the set of propositions $\{\theta_x : \theta \in LE\}$. Suppose then that according to the probability

measure P the propositions $(L_1)_x$ and $(L_2)_x$ are independent for all $x \in \Omega$. In this case ν is a functional but not truth functional measure. For example,

$$\alpha_1 \equiv L_1 \wedge L_2 : \nu_{L_1 \wedge L_2}(x) = \nu_{L_1}(x) \times \nu_{L_2}(x),$$
$$\alpha_2 \equiv L_1 \wedge \neg L_2 : \nu_{L_1 \wedge \neg L_2}(x) = \nu_{L_1}(x) \times (1 - \nu_{L_2}(x))$$
$$\alpha_3 = \neg L_1 \wedge L_2 : \nu_{\neg L_1 \wedge L_2}(x) = (1 - \nu_{L_1}(x)) \times \nu_{L_2}(x)$$
$$\alpha_4 = \neg L_1 \wedge \neg L_2 : \nu_{\neg L_1 \wedge \neg L_2}(x) = (1 - \nu_{L_1}(x)) \times (1 - \nu_{L_2}(x))$$

However, since ν is defined by a probability measure P then

$$\nu_{L_1 \wedge L_1}(x) = \nu_{L_1}(x) \neq \nu_{L_1}(x) \times \nu_{L_1}(x)$$

except when $\nu_{L_1}(x) = 0$ or $\nu_{L_1}(x) = 1$

  Clearly, however, $\nu_{L_1 \wedge L_1}(x)$ can be determined directly from $\nu_{L_1}(x)$ and $\nu_{L_2}(x)$ according to the function $f_{L_1 \wedge L_1}(a, b) = a$. Indeed $\nu_\theta(x)$, for any compound expression θ, can be evaluated recursively from $\nu_{L_1}(x)$ and $\nu_{L_2}(x)$ as a unique linear combinations of $\nu_{\alpha_i}(x) : i = 1, \ldots, 4$. For instance,

$$\nu_{L_1 \wedge L_1}(x) = \nu_{L_1 \wedge L_2}(x) + \nu_{L_1 \wedge \neg L_2}(x) =$$
$$\nu_{L_1}(x) \times \nu_{L_2}(x) + \nu_{L_1}(x) \times (1 - \nu_{L_2}(x)) = \nu_{L_1}(x)$$

In General

$$\nu_\theta(x) = \sum_{\alpha_i : \alpha_i \to \theta} \nu_{\alpha_i}(x)$$

Hence, we have that

$$\nu_{\theta \wedge \theta}(x) = \sum_{\alpha_i : \alpha_i \to \theta \wedge \theta} \nu_{\alpha_i}(x) = \sum_{\alpha_i : \alpha_i \to \theta} \nu_{\alpha_i}(x) = \nu_\theta(x) \text{ and}$$
$$\nu_{\theta \vee \neg \theta}(x) = \sum_{\alpha_i : \alpha_i \to \theta \vee \neg \theta} \nu_{\alpha_i}(x) = \sum_{i=1}^{4} \nu_{\alpha_i}(x) = 1$$

Clearly then ν satisfies idempotence and the law of excluded middle, and hence functional calculi are not in general subject to the restrictions of Dubois and Prade's theorem [20]

2.3   Operational Semantics for Membership Functions

In [103] Walley proposes a number of properties that any measure should satisfy if it is to provide an effective means of modelling uncertainty in intelligent systems. These include the following interpretability requirement:

> 'the measure should have a clear interpretation that is sufficiently defined to guide assessment, to understand the conclusions of the system and use them as a basis for action, and to support the rules for combining and updating measures'

Thus according to Walley an operational semantics should not only provide a means of understanding the numerical levels of uncertainty associated with propositions but must also provide some justification for the underlying calculus. In the case of fuzzy logic [110] this means than any interpretation of membership functions should be consistent with truth-functionality. If this turns out not to be the case then it may be fruitful to investigate new calculi for combining imprecise concepts.

In [25] Dubois and Prade suggest three possible semantics for fuzzy logic. One of these is based on the measure of similarity between elements and prototypes of the concept, while two are probabilistic in nature. In this section we shall review all three semantics and discuss their consistency with the truth-functionality assumption of fuzzy logic. We will also describe a semantics based on the risk associated with making an assertion involving vague concepts (Giles [34]).

2.3.1 Prototype Semantics

A direct link between membership functions and similarity measures has been proposed by a number of authors including Ruspini [89] and Dubois and Prade [25], [28]. The basic idea of this semantics is a follows: For any concept θ it is assumed that there a set of prototypical instances drawn from the universe Ω of which there is no doubt that they satisfy θ. Let $\mathcal{P}_\theta$ denote this set of prototypes for θ. It is also assumed that You have some measure of similarity according to which elements of the domain can be compared. Typically this is assumed to be a function $S : \Omega^2 \to [0,1]$ satisfying the following properties:

S1 $\forall x, y \in \Omega\ S(x,y) = S(y,x)$

S2 $\forall x \in \Omega\ S(x,x) = 1$ [4]

The membership function for the concept is then defined to be a subjective measure of the similarity between an element x and the closest prototypical element from $\mathcal{P}_\theta$:

$$\forall x \in \Omega\ \chi_\theta(x) = \sup\{S(x,y) : y \in \mathcal{P}_\theta\}$$

Clearly then if $x \in \mathcal{P}_\theta$ then $\chi_\theta(x) = 1$ and hence if $\mathcal{P}_\theta \neq \emptyset$ then $\sup\{\chi_\theta(x) : x \in \Omega\} = 1$. Also, if $\mathcal{P}_\theta = \emptyset$ then $\forall x \in \Omega\ \chi_\theta(x) = 0$ and hence according to prototype semantics all non-contradictory concepts have normalised membership functions.

We now consider the type of calculus for membership functions that could be consistent with prototype semantics. Clearly this can be reduced to the problem of deciding what relationships hold between the prototypes of concepts generated as combinations of more fundamental concepts and the prototypes of the component concepts. In other words, what are the relationships between $\mathcal{P}_{\neg\theta}$ and $\mathcal{P}_\theta$, between $\mathcal{P}_{\theta \wedge \varphi}$ and $\mathcal{P}_\theta$ and $\mathcal{P}_\varphi$, and between $\mathcal{P}_{\theta \vee \varphi}$ and $\mathcal{P}_\theta$ and P_φ.

For the case of $\neg\theta$ it would seem uncontroversial to assume that $\mathcal{P}_{\neg\theta} \subseteq (\mathcal{P}_\theta)^c$. Clearly, a prototypical *not tall* person cannot also be a prototypical *tall* person. In general, however, it would not seem intuitive to assume that $\mathcal{P}_{\neg\theta} = (P_\theta)^c$ since, for example, someone who is not prototypically *tall* may not necessarily be prototypically *not tall*.

For conjunctions of concepts one might naively assume that the prototypes for $\theta \wedge \varphi$ might correspond to the intersection $\mathcal{P}_\theta \cap \mathcal{P}_\varphi$. In this case it can easily be seen that:

$$\forall x \in \Omega \ \chi_{\theta \wedge \varphi}(x) \leq \min\left(\chi_\theta(x), \chi_\varphi(x)\right)$$

However, on reflection we might wonder whether a typical *tall and medium* person would be either prototypically *tall* or prototypically *medium*. This is essentially the basis of the objection to prototype theory (as based on Zadeh's min-max calculus) raised by Osherson etal. [75]. For example, they note that when considering the concepts *pet* and *fish* then a guppie is much more prototypical of the conjunction *pet fish* than it is of either of the conjuncts. Interestingly when viewed at the membership function level this suggests that the conjunctive combination of membership functions should not be monotonic (as it is for t-norms) since we would intuitively expect guppie to have a higher membership in the extension of *pet fish* than in either of the extensions of *pet* or *fish*.

In the case of disjunctions of concepts it does seem rather more intuitive that $\mathcal{P}_{\theta \vee \varphi} = \mathcal{P}_\theta \cup \mathcal{P}_\varphi$. For example, the prototypical *happy or sad* people might reasonably be thought to be composed of the prototypically *happy* people together with the prototypically *sad* people. In this case we obtain the strict equality:

$$\forall x \in \Omega \ \chi_{\theta \vee \varphi}(x) = \max\left(\chi_\theta(x), \chi_\varphi(x)\right)$$

Osherson etal. [75] argue against the use of prototypes to model disjunctions using a counter example based on the concepts *wealth*, *liquidity* and *investment*. The argument presented in [75] assumes that *wealth* corresponds to *liquidity or investment*, however, while there is certainly a relationship between these concepts it is not at all clear that it is a disjunctive one.

EXAMPLE 10 *Suppose the universe Ω is composed of five people:*

$$\Omega = \{Bill(B), \ Fred(F), \ Mary(M), \ Ethel(E), \ John(J)\}$$

with the following similarity measure S

$S(x, y)$	B	F	M	E	J
B	1	0.8	0.2	0.8	0.7
F	0.8	1	0.5	0.1	0.6
M	0.2	0.5	1	0.9	0.6
E	0.8	0.1	0.9	1	0.3
J	0.7	0.6	0.6	0.3	1

Now let

$$\mathcal{P}_{tall} = \{B, F\}\ \mathcal{P}_{medium} = \{M\}$$
$$\mathcal{P}_{\neg tall} = \{J\}, \mathcal{P}_{\neg medium} = \{B\}$$

The membership functions for tall, medium, not tall and not medium are then determined as follows:

$$\chi_{tall}(B) = \chi_{tall}(F) = 1$$
$$\chi_{tall}(M) = \max(S(M, B), S(M, F)) = \max(0.2, 0.5) = 0.5$$
$$\chi_{tall}(E) = \max(S(E, B), S(E, F)) = \max(0.8, 0.1) = 0.8$$
$$\chi_{tall}(J) = \max(S(J, B), S(J, F)) = \max(0.7, 0.6) = 0.7$$

$$\chi_{medium}(B) = S(B, M) = 0.2,\ \chi_{medium}(F) = S(F, M) = 0.5$$
$$\chi_{medium}(M) = 1,\ \chi_{medium}(E) = S(E, M) = 0.9$$
$$\chi_{medium}(J) = S(J, M) = 0.6$$

$$\chi_{\neg tall}(B) = S(B, J) = 0.7,\ \chi_{\neg tall}(F) = S(F, J) = 0.6$$
$$\chi_{\neg tall}(M) = S(M, J) = 0.6,\ \chi_{\neg tall}(E) = S(E, J) = 0.3$$
$$\chi_{\neg tall}(J) = 1$$

$$\chi_{\neg medium}(B) = 1,\ \chi_{\neg medium}(F) = S(F, B) = 0.8$$
$$\chi_{\neg medium}(M) = S(M, B) = 0.2,\ \chi_{\neg medium}(E) = S(E, B) = 0.8$$
$$\chi_{\neg medium}(J) = S(J, B) = 0.7$$

Now $\mathcal{P}_{tall \lor medium} = \mathcal{P}_{tall} \cup \mathcal{P}_{medium} = \{B, F, M\}$ *so that:*

$$\chi_{tall \lor medium}(B) = 1,\ \chi_{tall \lor medium}(M) = 1,\ \chi_{tall \lor medium}(F) = 1$$
$$\chi_{tall \lor medium}(E) = \max(S(E, B), S(E, F), S(E, M))$$
$$= \max(0.8, 0.1, 0.9) = 0.9,$$
$$\chi_{tall \lor medium}(J) = \max(S(J, B), S(J, F), S(J, M))$$
$$= \max(0.7, 0.6, 0.6) = 0.7$$

Clearly then

$$\forall x \in \Omega \; \chi_{tall \lor medium}(x) = \max(\chi_{tall}(x), \chi_{medium}(x))$$

Now $\mathcal{P}_{tall} \cap \mathcal{P}_{medium} = \emptyset$ *and hence if* $\mathcal{P}_{tall \land medium} = \mathcal{P}_{tall} \cap \mathcal{P}_{medium}$ *then*

$$\forall x \in \Omega \; \chi_{tall \land medium}(x) = 0$$

Alternatively, we might expect that Ethel would be a prototypical tall∧medium person since she has high membership both in tall and medium. Taking $\mathcal{P}_{tall \land medium} = \{E\}$ *gives:*

$$\chi_{tall \land medium}(B) = S(B, E) = 0.8, \; \chi_{tall \land medium}(F) = S(F, E) = 0.1,$$
$$\chi_{tall \land medium}(M) = S(M, E) = 0.9, \; \chi_{tall \land medium}(E) = 1,$$
$$\chi_{tall \land medium}(J) = S(J, E) = 0.3$$

What seems clear from the above discussion and example is that membership functions based on similarity measures are almost certainly not truth-functional and probably not even functional. For instance, the precise relationship between $\mathcal{P}_{\neg\theta}$ and $\mathcal{P}_\theta$ and between $\mathcal{P}_{\theta \land \varphi}$ and $\mathcal{P}_\theta$ and P_φ is problem specific. In other words, such relationships are strongly dependent on the meanings of θ and φ. It is the case, however, that if we take the prototypes of disjunctions as corresponding to the union of prototypes of the disjuncts then the resulting calculus will be partially-functional.

The fundamental problem with similarity semantics, however, is not principally related to the functionality of the emergent calculus. Rather it lies with the notion of similarity itself. In some concepts it may be straightforward to define the level of similarity between objects and prototypes. For example, in the case of the concept *tall* we might reasonably measure the similarity between individuals in terms of some monotonically decreasing function of the Euclidean distance between their heights. For other concepts, however, the exact nature of the underlying similarity measure would seem much harder to identify. For instance, consider the concept 'tree'. Now supposing we could identify a set of prototypical trees, itself a difficult task, how could we quantify the similarity between a variety of different plants and the elements of this prototype set? It is hard to identify an easily measurable attribute of plants that could be used to measure the degree of 'treeness'. To put it bluntly, the degree of similarity would seem as hard to define as the degree of membership itself!

2.3.2   Risk/Betting Semantics

Risk or betting semantics was proposed by Giles in a series of papers including [34] and [35]. In this semantics the fundamental idea (as described in [34]) is that the membership $\chi_\theta(x)$ quantifies the level of risk You are taking when You assert that 'x is θ'. Formally, let θ_x denote the proposition 'x is θ' then by asserting θ_x You are effectively saying that You will pay an opponent 1 unit if θ_x is false. Given this gamble we may then suppose that You associate a risk value $\langle \theta_x \rangle \in [0, 1]$ with the assertion θ_x. Such values are likely to be subjective probabilities and vary between agents, however, if $\langle \theta_x \rangle = 0$ we may reasonably suppose that You will be willing to assert θ_x and if $\langle \theta_x \rangle = 1$ You will certainly be unwilling to make such an assertion. We then define Your membership degree of x in θ as:

$$\chi_\theta (x) = 1 - \langle \theta_x \rangle$$

this being the subjective probability that θ_x is true. The standard connectives are then interpreted as follows:

Negation: You asserting $\neg \theta_x$ means that You will pay your opponent 1 unit if they will assert θ_x.

In this case You will be willing to assert $\neg \theta_x$ if You are sufficiently sure that θ_x is false, since in this case You will be repaid Your 1 unit by Your opponent. Hence, Your risk when asserting $\neg \theta_x$ is equivalent to one minus Your risk when asserting θ_x. Correspondingly,

$$\chi_{\neg \theta} (x) = 1 - \chi_\theta (x)$$

Conjunction: You asserting $(\theta \wedge \varphi)_x$ means that You agree to assert θ_x or φ_x, where the choice is made by Your opponent.

In this case the risk of asserting $(\theta \wedge \varphi)_x$ is the maximum of the risks associated with asserting θ_x and φ_x, since You have no way of knowing which of these two options Your opponent will choose. Hence, given the defined relationship between membership and risk we have

$$\chi_{\theta \wedge \varphi} (x) = \min (\chi_\theta (x), \chi_\varphi (x))$$

Disjunction: You asserting $(\theta \vee \varphi)_x$ means that You agree to assert either θ_x or φ_x, where the choice is made by You.

In this case, since You have the choice of which of the two statements to assert it is rational for You to choose the statement with minimum associated risk. In other words, $\langle (\theta \vee \varphi)_x \rangle$ should correspond to $\min (\langle \theta_x \rangle, \langle \varphi_x \rangle)$. Hence, the corresponding membership degree will satisfy:

$$\chi_{\theta \vee \varphi} (x) = \max (\chi_\theta (x), \chi_\varphi (x))$$

Although this semantics captures the fuzzy set calculus proposed by Zadeh [110], there is something odd about the truth-functional way in which the risks associated with asserting a compound expression are calculated. For instance, the rule for evaluating the risk of conjunction seems to implicitly assume that there is no logical relationship between the conjuncts. To see this consider the contradictory assertion $(\theta \wedge \neg\theta)_x$. In betting semantics this means that You are willing to either pay one unit to your opponent if θ_x is false (bet 1), or to pay Your opponent one unit if they will assert θ_x (bet 2). The choice between these two bets is then made by Your opponent. The main problem with this assertion is that given an informed opponent You are certain to lose. For instance, assuming Your opponent knows the truth value of θ_x then they would choose between bets 1 and 2 accordingly, as follows: If θ_x is true then Your opponent will pick bet 2 and You will lose 1 unit. Alternatively, if θ_x is false then Your opponent will pick bet 1 and You will lose 1 unit. Hence, given an informed opponent with knowledge of the truth value of θ_x, You are certain to lose 1 unit. In fact even when there is some doubt regarding the judgement of Your opponent You will tend to lose on average. For instance, if p is the probability that Your opponent will be correct regarding the truth value of θ_x then Your expected loss will be $p \times 1 + (1-p) \times 0 = p$. This suggests that Your actual risk when asserting $(\theta \wedge \neg\theta)_x$ is directly related to p and relatively independent of $\langle \theta_x \rangle$. Furthermore, since You will expect to lose in all cases except when $p = 0$, this being when Your opponent is definitely going to be wrong regarding θ_x, it is hard to imagine a scenario in which You would ever assert such a contradiction.

This problem with assuming a truth-functional calculus seems to have been recognised by Giles in later work [35] where he related the degree of membership $\chi_\theta(x)$ to a more general utility function $h_\theta(x)$. The latter is intended to quantify the degree of utility, possibly negative, which You will receive on asserting θ_x. There is no reason, however, why such a utility measure should be truth-functional, a fact highlighted by Giles [35] who comments that his research suggests 'that there is no viable truth-functional representation for conjunction and disjunction of fuzzy sentences'.

2.3.3   Probabilistic Semantics

Of the three semantics proposed by Dubois and Prade [25] two are probabilistic in nature. However, they do not offer a naive interpretation of fuzzy membership functions by, for example, claiming that they are simply probability distributions or density function that have not been normalised (e.g. Laviolette et al. [58] propose modelling linguistic concepts such as 'medium' and 'fast' using probability density functions). Rather they aim to model vague concepts in terms of the underlying uncertainty or variance associated with their meaning. In this section we discuss three such probabilistic semantics. This will include an in depth examination of random set semantics as mentioned by Dubois and Prade in [25] and developed at length by Goodman and Nguyen in a number of articles including [36], [37], [38], [72] and [73]. This work will be related to the voting model ([9], [32] and [59]) and the context model ([33] and [47]). We would ask for the reader's indulgence with respect to this extended discussion of random set interpretations of membership functions on the grounds that it is closely related to the semantics proposed subsequently in this volume. In addition, we will outline the likelihood interpretation of fuzzy sets as suggested by Hisdal [44] and show that, as indicated by Dubois and Prade [25], it is strongly related to the random set approach.

2.3.3.1   Random Set Semantics

In essence random sets are set valued variables and hence can be defined in terms of a mapping between two measurable domains as follows:

DEFINITION 11 *Let $\mathcal{B}$ be a σ-algebra on a universe U_1 and let $\mathcal{A}$ be a σ-algebra on 2^{U_2}, the power set of a second universe U_2. Then R is a random set from $\langle U_1, \mathcal{B} \rangle$ into $\langle 2^{U_2}, \mathcal{A} \rangle$ if $R : U_1 \rightarrow 2^{U_2}$ is a $\mathcal{B} - \mathcal{A}$ measurable function.*

If U_2 is finite and P is a probability measure on $\mathcal{B}$ then we can define a mass assignment (i.e. a probability distribution for R) according to:

$$\forall T \subseteq U_2 \; m(T) = P(R = T) = P(\{z \in U_1 : R(z) = T\})$$

This then generates a probability measure $\mathcal{M}$ on $\mathcal{A}$ as follows:

$$\forall A \in \mathcal{A} \; \mathcal{M}(A) = P(R \in A) = \sum_{T \in A} m(T)$$

DEFINITION 12 *Let R be a random set into 2^{U_2} then the fixed (or single) point coverage function is a function $cf_R : U_2 \rightarrow [0,1]$ such that:*

$$\forall w \in U_2 \; cf_R(w) = P(w \in R) = \sum_{T \subseteq U_2 : w \in T} m(T)$$

EXAMPLE 13 *Let $U_1 = \{a_1, a_2, a_3, a_4, a_5\}$ and $U_2 = \{b_1, b_2, b_3\}$, and let $\mathcal{B} = 2^{U_1}$ and $\mathcal{A} = 2^{2^{U_1}}$. Also, let P be defined according to the following values on the singleton sets:*

$$P(a_1) = 0.1, \; P(a_2) = 0.4, \; P(a_3) = 0.2, \; P(a_4) = 0.1, \; P(a_5) = 0.2$$

A possible random set from $\langle U_1, \mathcal{B} \rangle$ into $\langle 2^{U_2}, \mathcal{A} \rangle$ is:

$$R(a_1) = \{b_1, b_2\}, \; R(a_2) = \{b_2, b_3\}, \; R(a_3) = \{b_1, b_2, b_3\},$$
$$R(a_4) = \{b_1, b_2, b_3\}, R(a_5) = \{b_2, b_3\}$$

The corresponding mass assignment is given by:

$$m\left(\{b_1, b_2\}\right) = P\left(a_1\right) = 0.1, \ m\left(\{b_2, b_3\}\right) = P\left(a_2\right) + P\left(a_5\right)$$
$$= 0.4 + 0.2 = 0.6, m\left(\{b_1, b_2, b_3\}\right) = P\left(a_3\right) + P\left(a_4\right) = 0.2 + 0.1 = 0.3$$

The fixed point coverage function can then be evaluated as follows:

$$cf_R\left(b_1\right) = m\left(\{b_1, b_2\}\right) + m\left(\{b_1, b_2, b_3\}\right) = 0.1 + 0.3 = 0.4$$
$$cf_R\left(b_2\right) = m\left(\{b_1, b_2\}\right) + m\left(\{b_2, b_3\}\right) + m\left(\{b_1, b_2, b_3\}\right) = 1$$
$$cf_R\left(b_3\right) = m\left(\{b_2, b_3\}\right) + m\left(\{b_1, b_2, b_3\}\right) = 0.6 + 0.3 = 0.9$$

It is interesting to note that although any given mass assignment on 2^{U_2} yields a unique coverage function the converse does not hold. That is, there is generally a (sometimes infinite) set of mass assignments with the same fixed point coverage function. For instance, consider the coverage function given in example 13. What other mass assignments also have this coverage function? Since $cf_R\left(b_2\right) = 1$ it follows that, for any such mass assignment, $m\left(T\right) = 0$ for any subset T of U_2 not containing b_2. Now let:

$$m\left(\{b_1, b_2, b_3\}\right) = m_1, \ m\left(\{b_1, b_2\}\right) = m_2, \ m\left(\{b_2, b_3\}\right) = m_3$$
$$\text{and } m\left(\{b_2\}\right) = m_4$$

Now from the equation for the fixed point coverage function given in definition 12 we have that:

$$m_1 + m_2 + m_3 + m_4 = 1, \ m_1 + m_2 = 0.4, \ m_1 + m_3 = 0.9 \text{ so that}$$
$$m_2 = 0.4 - m_1, \ m_3 = 0.9 - m_1 \text{ and } m_4 = m_1 - 0.3 \text{ where } m_1 \in [0.3, 0.4]$$

Clearly, the mass assignment in example 13 corresponds to the case where $m_1 = 0.3$. Another, interesting case is where $m_1 = 0.4$ giving the mass assignment:

$$m\left(\{b_1, b_2, b_3\}\right) = 0.4, \ m\left(\{b_2, b_3\}\right) = 0.5, \ m\left(\{b_2\}\right) = 0.1$$

In this case where the subsets with non-zero mass form a nested hierarchy the underlying random set is referred to as consonant:

DEFINITION 14 *A consonant random set R into 2^{U_2} is such that*

$$\bigcup_{a \in U_1} R\left(a\right) = \{G_1, \ldots, G_t\} \text{ where } G_1 \subset G_2 \subset \ldots \subset G_t \subseteq U_2$$

In fact, as the following theorem shows, there is a unique consonant mass assignment consistent with any particular coverage function.

THEOREM 15 *Given a fixed point coverage function cf_R for which $\{cf_R(b) : b \in U_2\} = \{y_1, \ldots, y_n\}$ ordered such that $y_i > y_i + 1 : i = 1, \ldots, n - 1$ then any consonant random set with this coverage function must have the following mass assignment:*

For $F_i = \{b \in U_2 : cf_R(b) \geq y_i\}$

$m(F_n) = y_n$, $m(F_i) = y_i - y_{i+1} : i = 1, \ldots, n - 1$ *and* $m(\emptyset) = 1 - y_1$

Proof

Since R is consonant we have that

$$\{T : m(T) > 0\} = \{G_1, \ldots, G_t\} \text{ where } G_1 \subset G_2 \subset \ldots \subset G_t \subseteq U_2$$

Let $\overline{b_i} = \{b \in U_2 : cf_R(b) = y_i\}$ then there is a minimal value $i^* \in \{1, \ldots t\}$ such that $\overline{b_i} \subseteq G_{i^*}$ and $\overline{b_i} \cap G_j = \emptyset$ for $j < i^*$. Hence,

$$y_i = \sum_{k=i^*}^{t} m(G_k)$$

Now if $j < i$ then $\overline{b_j} \subseteq G_{i^*}$ since otherwise $j^* > i^*$ so that

$$y_j = \sum_{k=j^*}^{t} m(G_k) < \sum_{k=i^*}^{t} m(G_k) = y_i$$

which is a contradiction. Also for $j > i$ $\overline{b_j} \cap G_{i^*} = \emptyset$ since otherwise $j^* < i^*$ so that

$$y_j = \sum_{k=j^*}^{t} m(G_k) > \sum_{k=i^*}^{t} m(G_k) = y_i$$

which is a contradiction. From this is follows that:

$$G_{i^*} = \bigcup_{k=1}^{i} \overline{b_i} = \{b \in U_2 : cf_R(b) \geq y_i\}$$

Now it can easily be seem that $(i + 1)^* = i^* + 1$ hence

$$m(G_{i^*}) = \sum_{k=i^*}^{t} m(G_k) - \sum_{k=i^*+1}^{t} m(G_k) = y_i - y_{i+1} \text{ for } i = 1, \ldots n - 1$$

Now $G_{n^*} = \{b \in U_2 : cf_R(b) > 0\}$ and therefore $n = t$. From this it follows that*

$$y_n = m(G_t)$$

Also, suppose that $j < 1^*$ then either $G_j = \emptyset$ or $\exists b \in G_j$ such that $b \notin G_{1^*}$ which implies that $cf_R(b) > y_1$. This is a contradiction, so either $1^* = 1$ or $1^* = 2$ and $G_1 = \emptyset$. Hence, w.l.o.g and by relabelling the relevant sets we may assume that:

$$\{T : m(T) > 0\} = \{\emptyset, F_1, \ldots, F_n\} \text{ where } F_i = G_{i^*} : i = 1, \ldots, n$$

Finally, as required

$$m(\emptyset) = 1 - \sum_{k=1}^{n} m(F_k) = 1 - \left(\sum_{k=1}^{n-1} y_k - y_{k+1}\right) - y_n = 1 - y_1 \;\square$$

Another interesting mass assignment sharing the same coverage function as given in example 13 is obtained by setting $m_1 = 0.4 \times 0.9 \times 1 = 0.36$. This is the solution with maximum entropy and has the general form:

$$\forall T \subseteq U_2 \, m(T) = \left(\prod_{b \in T} cf_R(b)\right) \times \left(\prod_{b \notin T} (1 - cf_R(b))\right)$$

Essentially, the idea of random set semantics is that vague concepts are concepts for which the exact definition is uncertain. In this case the extension of a vague concept θ is defined by a random set R_θ into 2^Ω (i.e. $U_2 = \Omega$) with associated mass assignment m. The membership function $\chi_\theta(x)$ is then taken as coinciding with the fixed point coverage function of R_θ so that:

$$\forall x \in \Omega \, \chi_\theta(x) = cf_{R_\theta}(x) = P(x \in R_\theta) = \sum_{T \subseteq \Omega : x \in T} m(T)$$

Within the context of random set semantics, given that for a particular element of U_1 the extension of a concept is a crisp set of elements from Ω, it is perhaps natural to assume that the following classical laws hold when combining vague concepts:

$$\forall \theta \in LE \, \forall a \in U_1 \, R_{\neg\theta}(a) = R_\theta(a)^c$$
$$\forall \theta, \varphi \in LE \, \forall a \in U_1 \, R_{\theta \wedge \varphi}(a) = R_\theta(a) \cap R_\varphi(a)$$
$$\forall \theta, \varphi \in LE \, \forall a \in U_1 \, R_{\theta \vee \varphi}(a) = R_\theta(a) \cup R_\varphi(a)$$

In this case it can easily seen that the associated calculus of membership (fixed point coverage) functions will, in general, be neither functional or truth-functional. Certainly such a calculus would satisfy the standard Boolean laws including idempotence (since $R_\theta(a) \cap R_\theta(a) = R_\theta(a)$), the law of excluded middle (since $R_\theta(a) \cup R_\theta(a)^c = \Omega$) and the law of non-contradiction (since $R_\theta(a) \cap R_\theta(a)^c = \emptyset$).

2.3.3.2   Voting and Context Model Semantics

One aspect of the random set semantics that remains unclear from the above discussion is the exact nature of the uncertainty regarding the extension of θ. In more formal terms this corresponds to asking what is the exact nature of the universe U_1. One possibility is to assume that the random set uncertainty comes from the variation in the way that concepts are defined across a population. That is we take U_1 to be a population of individuals V each of whom are asked to provide an exact(crisp) extension of θ. Alternatively, for finite universes with small numbers of elements, these extensions may be determined implicitly by asking each individual whether of not they think that each element satisfies θ. This is the essence of the voting model for fuzzy sets proposed originally by Black [9] and later by Gaines [32]. Hence, in accordance with the random set model of fuzzy membership we have that

$$\chi_\theta(x) = P(\{v \in V : x \in R_\theta(v)\})$$

and assuming P is the uniform distribution then

$$= \frac{|\{v \in V : x \in R_\theta(v)\}|}{|V|}$$

Hence, when P is the uniform distribution on voters then $\chi_\theta(x)$ is simply the proportion of voters that include x in the extension of θ (or alternatively agree that 'x is θ'). Now assuming that voters apply the classical rules for the logical connectives then as discussed above such a random set based model will not be truth-functional. In fact it is well known that such measures of uncertainty defined in terms of relative frequency are probability measures (see Paris [78] for an exposition). As an alternative, Lawry [59] proposed a non-classical mechanism according to which voters could decide whether or not an element satisfied a concept as defined in terms of a logical combinations of labels. This mechanism corresponds to the following non-standard extension of the classical logic notion of a valuation.

DEFINITION 16 *A fuzzy valuation for instance* $x \in \Omega$ *is a function* $F_x :$ $LE \times [0,1] \rightarrow \{1,0\}$ *satisfying the following conditions (see figure 2.3):*

(i) *$\forall \theta \in LE$ and $\forall y, y' \in [0,1]$ such that $y \leq y'$ then $F_x(\theta, y) = 0 \Rightarrow$* $F_x(\theta, y') = 0$

(ii) *$\forall \theta, \varphi \in LE$ and $\forall y \in [0,1]$ $F_x(\theta \wedge \varphi, y) = 1$ iff $F_x(\theta, y) = 1$ and* $F_x(\varphi, y) = 1$ *(see figure 2.4)*

(iii) *$\forall \theta, \varphi \in LE$ and $\forall y \in [0,1]$ $F_x(\theta \vee \varphi, y) = 1$ iff $F_x(\theta, y) = 1$ or* $F_x(\varphi, y) = 1$

(iv) *$\forall \theta \in LE$ and $\forall y \in [0,1]$ $F_x(\neg\theta, y) = 1$ iff $F_x(\theta, 1-y) = 0$ (see figure* 2.5)

Fuzzy valuations are an extension of classical valuations where the truth value assigned is dependent not only on the expression but also on a parameter y between 0 and 1 representing the degree of scepticism of the voter. The closer y is to 0 the less sceptical the voter and the more likely they are to be convinced of the truth of any given statement (i.e. to assign a truth-value of 1) and conversely the closer y is to 1 the more sceptical the voter and the more likely it is that they will not be convinced of the truth of the expression (i.e. to assign a truth-value of 0). The scepticism level, then, should be thought of as representing an internal state of the agent according to which their behaviour is more of less cautious. Fuzzy valuations attempt to capture formally the following description of voter behaviour as given by Gaines [32]:

> 'members of the population each evaluated the question according to the same criteria but applied a different threshold to the resulting evidence, or 'feeling'. The member with the lowest threshold would then always respond with a yes answer when any other member did, and so on up the scale of thresholds.'

Now assuming y varies across voters we may suppose that the probability distribution on scepticism levels is given by a probability measure ρ on the Borel subsets of $[0, 1]$. Given this measure, the membership degree $\chi_\theta(x)$ relative to a fuzzy valuation F_x can then be defined as the probability of that a voter has a scepticism level y such that $F_x(\theta, y) = 1$. More formally,

DEFINITION 17 *Given a fuzzy valuation F_x we can define a corresponding membership function for any $\theta \in LE$ according to:*

$$\chi_\theta(x) = \rho(\{y \in [0, 1] : F_x(\theta, y) = 1\})$$

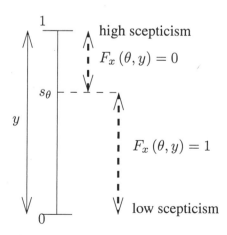

Figure 2.3: Diagram showing how fuzzy valuation F_x varies with scepticism level y

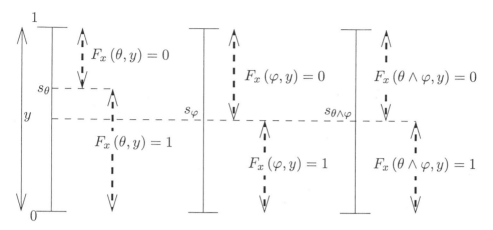

Figure 2.4: Diagram showing the rule for evaluating the fuzzy valuation of a conjunction at varying levels of scepticism

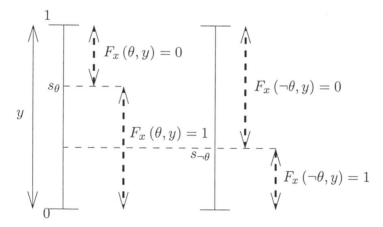

Figure 2.5: Diagram showing the rule for evaluating the fuzzy valuation of a negation at varying levels of scepticism

THEOREM 18 *For membership degrees based on fuzzy valuation F_x we have that*

$$\forall \theta, \varphi \in LE \; \chi_{\theta \wedge \varphi}(x) = \min(\chi_\theta(x), \chi_\varphi(x))$$

Proof

$\chi_{\theta \wedge \varphi}(x) = \rho(\{y \in [0,1] : F_x(\theta, y) = 1, F_x(\varphi, y) = 1\})$ *and*
$\{y \in [0,1] : F_x(\theta, y) = 1, F_x(\varphi, y) = 1\} =$
$\{y \in [0,1] : F_x(\theta, y) = 1\} \cap \{y \in [0,1] : F_x(\varphi, y) = 1\}$

Now $\forall \theta \in LE$ let

$s_\theta = \sup \{y \in [0,1] : F_x(\theta, y) = 1\}$ *so that either*
$\{y \in [0,1] : F_x(\theta, y) = 1\} = [0, s_\theta]$ *or* $= [0, s_\theta)$

Then w.l.o.g suppose that $\chi_\varphi(x) \leq \chi_\theta(x)$ which implies that $s_\varphi \leq s_\theta$. In this case (see figure 2.4),

$\{y \in [0,1] : F_x(\theta, y) = 1\} \cap \{y \in [0,1] : F_x(\varphi, y) = 1\} =$ *either*
$[0, s_\theta] \cap [0, s_\varphi] = [0, s_\varphi]$ *or* $[0, s_\theta] \cap [0, s_\varphi) = [0, s_\varphi)$ *or*
$[0, s_\theta) \cap [0, s_\varphi] = [0, s_\varphi]$ *or* $[0, s_\theta) \cap [0, s_\varphi) = [0, s_\varphi)$

In all of these cases

$\{y \in [0,1] : F_x(\theta, y) = 1, F_x(\varphi, y) = 1\} = \{y \in [0,1] : F_x(\varphi, y) = 1\}$

and hence $\chi_{\theta \wedge \varphi}(x) = \chi_\varphi(x)$ as required. $\square$

THEOREM 19 *For membership degrees based on fuzzy valuation F_x we have that*

$\forall \theta, \varphi \in LE \ \chi_{\theta \vee \varphi}(x) = \max(\chi_\theta(x), \chi_\varphi(x))$

Proof
Follows similar lines to theorem 18. $\square$

THEOREM 20 *Provided that ρ is a symmetric probability measure satisfying $\forall a, b \in [0,1] \ \rho([a,b]) = \rho([1-b, 1-a])$ then for membership degrees based on fuzzy valuation F_x we have that*

$\forall \theta \in LE \ \chi_{\neg \theta}(x) = 1 - \chi_\theta(x)$

Proof
By the symmetry condition on ρ it follows that $\forall a \in [0,1] \ \rho([0,a]) = \rho([1-a, 1])$ and $\rho([0,a)) = \rho((1-a, 1])$. Also recall that either $\{y \in [0,1] : F_x(\theta, y) = 1\} = [0, s_\theta]$ or $= [0, s_\theta)$.
Now

$\chi_{\neg \theta}(x) = \rho(\{y \in [0,1] : F_x(\neg \theta, y) = 1\}) =$ *by definition 16 part (iv)*
$\rho(\{y \in [0,1] : F_x(\theta, 1-y) = 0\})$

Supposing that

$\{y \in [0,1] : F_x(\theta, y) = 1\} = [0, s_\theta]$ *then by definition 16 part (iv)*
$F_x(\neg \theta, y) = 1$ *iff $1 - y > s_\theta$ iff $y < 1 - s_\theta$ and therefore (see figure 2.5)*
$\{y \in [0,1] : F_x(\neg \theta, y) = 1\} = [0, 1 - s_\theta) = [1 - s_\theta, 1]^c$ *therefore*
$\rho(\{y \in [0,1] : F_x(\neg \theta, y) = 1\}) = 1 - \rho([1 - s_\theta, 1]) = 1 - \rho([0, s_\theta])$
$= 1 - \chi_\theta(x)$ *by the symmetry condition on ρ*

Alternatively supposing that

$\{y \in [0,1] : F_x(\theta, y) = 1\} = [0, s_\theta)$ *then*

$\{y \in [0,1] : F_x(\neg\theta, y) = 1\} = [0, 1 - s_\theta] = (1 - s_\theta, 1]^c$ *and therefore*

$\rho(\{y \in [0,1] : F_x(\neg\theta, y) = 1\}) = 1 - \rho((1 - s_\theta, 1]) = 1 - \rho([0, s_\theta))$

$= 1 - \chi_\theta(x) \ \square$

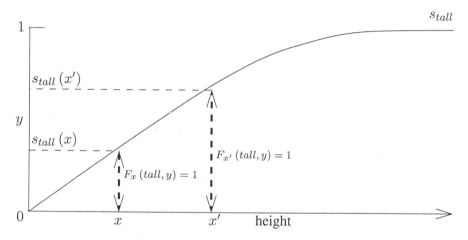

Figure 2.6: Diagram showing how the range of scepticism values for which an individual is considered tall increases with height

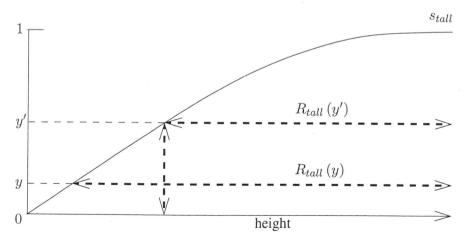

Figure 2.7: Diagram showing how the extension of tall varies with the y

Now as for each $x \in \Omega$ it is assumed that every voter defines a distinct fuzzy valuation F_x across which the upper bound on scepticism levels for which θ is true, s_θ, will vary. Hence, we can view $s_\theta : \Omega \rightarrow [0, 1]$ as a function of x.

For example, we might expect the value of s_{tall} to be an increasing function of height since the greater the height of an individual the more likely a voter will be to agree that this individual is *tall* even if that voter has a relatively high scepticsm level (see figure 2.6). Also, given such a set of fuzzy valuations then for each level y we can naturally generate the extension of a concept θ by considering all elements x for which $F_x(\theta, y) = 1$ (see figure 2.7). This motivates the following definition:

DEFINITION 21 *Given a set of fuzzy valuations F_x for every $x \in \Omega$ then for any $\theta \in LA$ the extension of θ is naturally defined by the function $R_\theta : [0,1] \to 2^{U_2}$ as follows:*

$$\forall y \in [0,1] \; R_\theta(y) = \{x \in \Omega : F_x(\theta, y) = 1\}$$

The extension of a concept defined in this way, relative to a set of fuzzy valuations, corresponds to a consonant random set as shown by the following theorem.

THEOREM 22 *For any $\theta \in LE$, given a set of fuzzy valuations F_x for every $x \in \Omega$, then R_θ as defined in definition 21 is consonant random set from $\langle [0,1], \mathcal{B} \rangle$ into $\left\langle 2^\Omega, 2^{2^\Omega} \right\rangle$ where $\mathcal{B}$ are the Borel subsets of $[0,1]$ and (for mathematical simplicity) it is assumed that Ω is finite.*
  Proof
For $T \subseteq \Omega \; R_\theta^{-1}(T) = \{y \in [0,1] : R_\theta(y) = T\}$ can be determined as follows: Let

$$\overline{s}_T = \min(s_\theta(x) : x \in T) \text{ and } \underline{s}_T = \max(s_\theta(x) : x \notin T)$$

Now if $y \in R_\theta^{-1}(T)$ then $y \leq \overline{s}_T$ since otherwise $\exists x \in T$ such that $y > s_\theta(x)$ which implies $F_x(\theta, y) = 0$ and hence $x \notin R_\theta(y)$ which would mean that $R_\theta(y) \neq T$. Similarly, if $y \in R_\theta^{-1}(T)$ then $y \geq \underline{s}_T$ since otherwise $\exists x \in T^c$ such that $y < s_\theta(x)$ which implies $F_x(\theta, y) = 1$ and hence $x \in R_\theta(y)$ which would mean that $R_\theta(y) \neq T$. Also, if $y < \overline{s}_T$ and $y > \underline{s}_T$ then $\forall x \in T$ $F_x(\theta, y) = 1$ and $\forall x \in T^c$ $F_x(\theta, y) = 0$ so that $R_\theta(y) = T$. From this we can see that

$$R_\theta^{-1}(T) = \text{ either}$$
$$\emptyset \text{ (if } \overline{s}_T < \underline{s}_T) \text{ or } [\underline{s}_T, \overline{s}_T] \text{ or } [\underline{s}_T, \overline{s}_T) \text{ or } (\underline{s}_T, \overline{s}_T] \text{ or } (\underline{s}_T, \overline{s}_T)$$

All of these sets are Borel measureable as required.
We now show that R_θ is consonant. If $y' > y$ then by definition 16 part (i) it

follows that

$\{x \in \Omega : F_x(\theta, y) = 0\} \subseteq \{x \in \Omega : F_x(\theta, y') = 0\}$ *and therefore*
$\{x \in \Omega : F_x(\theta, y') = 0\}^c \subseteq \{x \in \Omega : F_x(\theta, y) = 0\}^c$ *which implies*
$\{x \in \Omega : F_x(\theta, y') = 1\} \subseteq \{x \in \Omega : F_x(\theta, y) = 1\}$ *and hence that*
$R_\theta(y') \subseteq R_\theta(y)$ *as required.* $\square$

THEOREM 23 *Given R_θ as defined in definition 21 and χ_θ as defined in definition 17 then*

$$\forall x \in \Omega \; \chi_\theta(x) = cf_{R_\theta}(x)$$

Proof

$$\forall x \in \Omega \; cf_{R_\theta}(x) = \sum_{S:x\in S} m(S) = \sum_{S:x\in S} \rho(\{y \in [0,1] : R_\theta(y) = S\})$$
$$= \rho(\{y \in [0,1] : x \in R_\theta(y)\}) = \rho(\{y \in [0,1] : F_x(\theta, y) = 1\}) = \chi_\theta(x) \; \square$$

While fuzzy valuations provide an interesting mechanism according to which a truth-functional calculus can emerge from a random set interpretation of membership functions the theory remains problematic with respect to its treatment of negation. Specifically the negation rule given in definition 16 part (iv) is hard to justify. Lawry [59] suggests that:

> 'In order to determine a truth value for $\neg\theta$ while in state y the agent converts to a dual state $1 - y$ to evaluate the truth value of θ. The truth value of $\neg\theta$ is then taken to be the opposite of this truth value for θ.'

This, however, would appear to be a somewhat convoluted method for evaluating the negation of an expression without a clear semantic justification. Paris [80] proposes the following alternative justification:

> 'if voters with a low degree of scepticism y reject θ (i.e. $F(\theta, y) = 0$ for some low value of y) then other voters, even those with a relatively high degree of scepticism, would see this as support for $\neg\theta$ and be influenced to vote accordingly'

The problem with this argument is that it presupposes that individual voters will have access not only to the voting response of other voters but also their scepticism level at the time of voting. This would seem a very unrealistic assumption.

In [33] Gebhardt and Kruse proposed a variant on the voting model in which the elements of U_1 are interpreted as different contexts across which vague concepts have different extensions . For example, in [47] it is suggested that in the case where $LA = \{very\ short,\ short,\ medium,\ ...\}$ the contexts (i.e. elements of U_1) might correspond to nationalities such as Japanese, American, Swede, etc. Once again, however, it is clear that such a calculus will not in general be functional and as stated in [47] it is only possible to restrict $\chi_{\theta \wedge \varphi}(x)$ to the standard probabilistic interval so that:

$$\forall x \in \Omega, \ \forall \theta, \varphi \in LE$$
$$\chi_{\theta \wedge \varphi}(x) \in [\max(0, \chi_\theta(x) + \chi_\varphi(x) - 1), \min(\chi_\theta(x), \chi_\varphi(x))]$$

Interestingly, [47] indentify a special case of the context model for which a limited form of functionality does exist. Here it is supposed that the total set of labels is comprised of two distinct subsets LA_1 and LA_2 with different sets of contexts (C_1 and C_2 respectively) so that $U_1 = C_1 \times C_2$. For example, LA_1 might be the set of height labels described above with C_1 the associated set of nationality contexts, while LA_2 might be a set of labels relating to income (e.g. *high*, *low*) and C_2 a set of residential areas. Now it is reasonable to assume that for such different sets of contexts the occurence of a particular context from C_1 will be independent of the occurrence of any context from C_2. Hence, if P_1 is the probability distribution on C_1 and P_2 is the probability distribution on C_2 then the probability distribution on the joint space of contexts U_1 will be $P_1 \times P_2$. In this case, if LE_1 is the set of expression generated from labels LA_1 and LE_2 is the set of expression generated from labels LA_2 then for $\theta \in LE_1$ and $\varphi \in LE_2$ is can easily be seen that:

$$\forall x \in \Omega \ \chi_{\theta \wedge \varphi}(x) = \chi_\theta(x) \times \chi_\varphi(x)$$
$$\forall x \in \Omega \ \chi_{\theta \vee \varphi}(x) = \chi_\theta(x) + \chi_\varphi(x) - \chi_\theta(x) \times \chi_\varphi(x)$$

2.3.3.3   Likelihood Semantics

Likelihood semantics was first proposed as part of the TEE (Threshold, Error, Equivalence) model by Hisdal [44] and, as point out by Dubois and Prade [25], is closely related to random set semantics. Hisdal [44] suggests that membership functions can be derived from so-called yes-no experiments where a population of individuals are asked whether or not a particular label expression θ can be used to describe a certain value x. The membership of x in the extension of θ is then taken to be defined as follows:

$$\chi_\theta(x) = P(\theta|x)$$

where the above probability corresponds to the likelihood that a randomly chosen individual will respond with a yes to the question 'is this value θ?' given that the value is x.

It is indeed not difficult to see connections between Hisdal's yes-no experiments and the voting model described above. For instance, we might assume that an individual (voter) will repond yes that θ can describe x if and only if x is an element of that individual's extension of θ generated as part of a voting experiment of the type discussed in the previous section. In this case

$$P(\theta|x) = P(\{v \in V : x \in R_\theta(v)\})$$

Given this relationship and indeed the fundamental properties of probability theory it is clear that fuzzy memberships based on likelihoods will not, in general, be truth-functional nor even functional.

Summary

In this chapter we have given an overview of the theory of fuzzy sets as proposed by Zadeh [111]. This theory identifies a formal truth-functional calculus for membership functions where the membership value of a compound fuzzy set is obtain by applying truth functions for the various connectives to the component membership functions. Truth-functionality was then compared with a weaker form of functionality in terms of the Boolean properties that can be satisfied by the resulting calculus.

The review of fuzzy set theory focused particularly on possible operational semantics for fuzzy membership functions and the consistency of each proposed semantics with Zadeh's truth-functional calculus was investigated. Indeed it is this semantic based analysis that highlights the real problem with truth-functional fuzzy set theory. For while a notion of membership function based on any of the interpretation discussed in this chapter may indeed prove to be a useful tool for modelling vague concepts none provide any convincing justification that the calculus for combining such membership function should be truth-functional. In other words, the real criticism of fuzzy logic is not that it fails to satisfy any particular Boolean property (as suggested by Elkan [30]) but rather that it has no operational semantics which is consistent with its truth-functional calculus. As such it fails to satisfy Walley's [103] interpretability principle for uncertainty measures.

There are two main responses to this criticism of fuzzy set theory. The first is that while no totally convincing semantics has been identified that justifies the truth-functionality assumption this does not mean that such a semantics will not be identified in the future. This is undeniably true but it will certainly necessitate that more attention is paid by the research community to the issue of operational semantics for membership functions than is currently the case. The second response is that the lack of a semantics does not matter since we can take membership values as primitives in the same way as crisp membership values are primitives. However, this position would seem somewhat hard to justify. For instance, unlike crisp sets there are no physical realisations of fuzzy sets. That is while there are crisp sets of objects that occur in the physical world, fuzzy sets, if they occur at all, can only really occur as subjective constructs within a certain linguistic context. For physical realisations of crisp sets membership functions are objective measurements of reality and it seems likely that it is from this connection to the physical world that many of our intuitions regarding the calculus for crisp memberships are derived. For fuzzy sets, even assuming truth-functionality based on the t-norm, t-conorm and negation axioms, there is no consensus regarding the definition of truth functions for membership values. Indeed, as we have seen, there is does not seem to be any intuitive justification for the truth-functionality assumption itself.

If membership functions for the extensions of vague concepts exist, or indeed even if they are 'convenient fictions' for modelling vagueness as suggested by Dubois and Prade [24], how can we evaluate them? Certainly we would not expect individuals to be able to reliably estimate their own fuzzy membership function for a vague concept through some process of introspection. Even if membership functions are in some way represented in an individual's head, as suggested by Hajek [42], there would be no reason to suppose that they would have access to them (as is admitted by Hajek). In this case we require some behavioural mechanism according to which we can elicit membership function from individuals, this in itself requiring a lower-level understanding of membership functions and the way they are used.

Notes

1 One is tempted to say that either 'x is θ' or that 'x is a θ' should be a declarative statement but this would perhaps prejudge any discussion on the allocation of truth values to fuzzy statements

2 alternatively full compositionality [22] or strong functionality [67]

3 Weak Functionality in Lawry [67]

4 In many cases the following generalization of the triangle inequality is also required as property: $\forall x, y, z \in \Omega \; S(x, z) \geq f_\wedge (S(x, y), S(y, z))$ for some t-norm $f_\wedge$.

Chapter 3

LABEL SEMANTICS

3.1 Introduction and Motivation

Most current approaches to modelling vague or imprecise concepts are based
on the explicit definition of the extension of that concept, corresponding to the
(fuzzy) set of objects for which the concept holds (see [117]). For example, in
a recent analysis of models of vagueness Dubois et al [28] categories different
semantics for fuzzy sets on the basis of the nature of the division between the
extension of a vague concept and its negation. It is unlikely, however, that human
reasoning with imprecise concepts is based on the explicit use of their extensions
irrespective of whether they are fuzzy or crisp. Indeed, for many concepts such
extensions are so diverse that they would require large amounts of memory even
to store. For example, for a particular individual the concept 'tree' would have
an extension containing all the species of tree (or possibly all the different trees)
that that individual has ever classified as 'a tree'. Certainly such a set would
include a large number and variety of objects. It is perhaps more likely that
individuals have some heuristic mechanism or rule for deciding whether or not
an object can be described as a tree under certain circumstances and conditions.
This mechanism might be a model describing certain shared attributes of trees
or perhaps a means of comparing a current object with previously encountered
prototypical trees. For example, the following definition of a tree taken from
the Oxford English Dictionary might be viewed as a linguistic description of
such a mechanism:

> 'A perennial plant having a self-supporting woody main stem or trunk (which usually
> develops woody branches at some distance from the ground), and growing to a consid-
> erable height and size. (Usually distinguished from a bush or shrub by size and manner
> of growth...)'

We would certainly expect this mechanism for classification to differ between
individuals across a population but we would also expect this difference to

be limited, assuming a shared linguistic and cultural context, since the concept must be useful for conveying information. Rohit Parikh [77] illustrates how two individuals with different notions of the colour 'blue' can still effectively use the concept to pass information between them by considering how the search space of one person looking for a book required by the other is reduced when they learn that the book is 'blue'. In the same article Parikh also advises an investigation into the use of vague terms as part of so called language games in order to obtain an 'anti-representational' view of vague predicates emphasising 'assertibility rather than truth'.

What cannot be denied is that humans posses a mechanism for deciding whether or not to make assertions (e.g. 'Bill is tall') or to agree to a classification (e.g. 'Yes, that is a tree'). Further, although the concepts concerned are vague this underlying decision process is fundamentally crisp (bivalent). For instance, You are either willing to assert that 'x is a *tree*' or You are not. In other words, either *tree* is an appropriate label to describe x or it is not. As an intelligent agent You are continually faced with making such crisp decisions regarding vague concepts as part of Your every day use of language. Of course, You may be uncertain about labels and even express these doubts (e.g. 'I'm not sure whether you would call that a tree or a bush') but the underlying decision is crisp. In this chapter we introduce label semantics as providing a new perspective on vague concepts by focusing on the decision process You must go through in order to identify which labels or expressions can actually be used to describe an object or instance. The foundation of this theory will be based on quantifying Your subjective belief that a label L is appropriate to describe an object x and hence whether or not You can reasonably assert that 'x is L'. Such belief evaluation would be made on the basis of Your previous experience of the use of labels and label expressions by other agents to describe other similar instances. As mentioned in chapter 1 such a procedure is only really meaningful if agents adopt an 'epistemic stance', whereby they assume that there exist certain conventions for the correct use of the labels in LA to describe the elements of Ω. But since a fundamental goal of language is to share information across a population this stance does not seem intuitively unreasonable. Of course, such a linguistic convention does not need to be imposed by some outside authority like the Oxford English Dictionary or the Academia Lengua Espanola, but instead would emerge as a result of interactions between agents each adopting the 'epistemic stance'. In the sequel we shall show that such a calculus can, in a restricted way, share some of the mathematical apparatus of fuzzy set theory but the underlying motivation and interpretation will remain distinct. We will also suggest that such a theory should not be truth-functional but might be functional.

3.2 Appropriateness Measures and Mass Assignments on Labels

A finite set of labels $LA = \{L_1, \ldots, L_n\}$ are identified as possible descriptions of the elements of a universe Ω. For any element $x \in \Omega$ You make the bivalence assumption that a label $L \in LA$ is either an appropriate description of x or it is not, however, You are uncertain which of these possibilities hold. Let $\mu_L(x) \in [0, 1]$ quantify Your degree of belief[1] that L is an appropriate label for x. $\mu_L(x)$ can be thought of as quantifying, what You judge to be the level of appropriateness of L as a label for x.

Given the bivalence assumption regarding the appropriateness of labels then You must recognise the existence of a crisp set of appropriate labels for x, denoted $\mathcal{D}_x$. In other words, $\mathcal{D}_x$ corresponds to the set of labels that are appropriate with which to describe x. Now since You are uncertain as to whether or not a label is appropriate with which to describe x You will also be uncertain of the exact definition of $\mathcal{D}_x$. For all $T \subseteq LA$ let $m_x(T) \in [0, 1]$ correspond to Your belief that $\mathcal{D}_x = T$. The measure m_x can be viewed as a mass assignment on labels defined as follows:

DEFINITION 24 *Mass Assignments on Labels*
A mass assignment on labels is a function $m : 2^{LA} \to [0, 1]$ *such that*
$\sum_{T \subseteq LA} m(T) = 1$

Notice that in definition 24 there is no requirement for the mass associated with the empty set to be zero. In the context of label semantics $m_x(\emptyset)$ quantifies Your belief that no labels are appropriate to describe x. We might observe that this phenomena occurs frequently in natural language especially when labelling perceptions generated along some continuum. For example, we occasionally encounter colours for which none of our available colour descriptors seem appropriate. Hence, the value $m_x(\emptyset)$ is an indicator of the describability of x in terms of the labels LA.

The appropriateness of a label L to x is clearly related to the mass assignment on labels m_x in the following sense. Your belief that L is an appropriate label for x should correspond to Your belief that the set of appropriate labels for X contains L. In other words, that

$$\forall L \in LA \; \forall x \in \Omega \; \mu_L(x) = \sum_{T \subseteq LA : L \in T} m_x(T)$$

3.3 Label Expressions and λ-Sets

As well as labels in LA, You may also want to evaluate the appropriateness of more complex logical expressions. For example, You may want to determine to what degree the expression *tall but not very tall* is appropriate as a description of a particular individual, where $tall, \; very \; tall \in LA$. Consider then the set of logical expressions LE obtained by recursive application of the standard logical connectives in LA. In order to evaluate the appropriateness of such expressions You must identify what information they provide regarding the appropriateness of labels in LA. For example, the expression *tall but not very tall* tells You that *tall* is appropriate but *very tall* is not. Expressed in terms of $\mathcal{D}_x$ then the meaning of *tall but not very tall* corresponds to the constraint $\mathcal{D}_x \in \{T \subseteq LA : tall \in T, very \; tall \notin T\}$. In general, for any label expression $\theta \in LE$ we should be able to identify a maximal set of label sets, $\lambda(\theta)$, that are consistent with θ so that the meaning of θ can be interpreted as the constraint $\mathcal{D}_x \in \lambda(\theta)$.

DEFINITION 25 *λ-Sets*
$\lambda : LE \to 2^{LA}$ *is defined recursively as follows:* $\forall \theta, \; \varphi \in LE$

(i) $\forall L_i \in LA \; \lambda(L_i) = \{T \subseteq LA : L_i \in T\}$

(ii) $\lambda(\theta \wedge \varphi) = \lambda(\theta) \cap \lambda(\varphi)$

(iii) $\lambda(\theta \vee \varphi) = \lambda(\theta) \cup \lambda(\varphi)$

(iv) $\lambda(\neg\theta) = \lambda(\theta)^c$

(v) $\lambda(\theta \to \varphi) = \lambda(\neg\theta) \cup \lambda(\varphi)$

EXAMPLE 26 *Let* $LA = \{small, medium, large\}$ *then*

$\lambda(small \wedge medium) = \{\{small, medium\}, \{small, medium, large\}\}$
$\lambda(small \vee medium) = \{\{small\}, \{medium\}, \{small, medium\},$
$\{small, large\}, \{medium, large\}, \{small, medium, large\}\}$
$\lambda(small \to medium) = \{\{small, medium\}, \{small, medium, large\},$
$\{medium, large\}, \{medium\}, \{large\}, \emptyset\}$
$\lambda(\neg small) = \{\{medium\}, \{large\}, \{medium, large\}, \emptyset\}$

Intuitively, $\lambda(\theta)$ corresponds to those subsets of LA identified as being possible values of $\mathcal{D}_x$ by expression θ. In this sense the imprecise linguistic restriction 'x is θ' on x corresponds to the strict constraint $\mathcal{D}_x \in \lambda(\theta)$ on $\mathcal{D}_x$. Hence, we can view label descriptions as an alternative to Zadeh's linguistic variables [112], [113], [114] as a means of encoding linguistic constraints. In this context a natural way for You to evaluate the appropriateness of $\theta \in LE$ is to aggregate the values of m_x across $\lambda(\theta)$. This motivates the following general definition of appropriateness measures.

DEFINITION 27 *Appropriateness Measures*
$\forall \theta \in LE, \; \forall x \in \Omega$ *the measure of appropriateness of* θ *as a description of* x

is given by:

$$\mu_\theta(x) = \sum_{T \in \lambda(\theta)} m_x(T)$$

3.4 A Voting Model for Label Semantics

In addition to this subjective interpretation of fuzzy labels an operational semantics can also be given in terms of a voting model. In this case a population of individuals is asked to identify a (crisp) set of appropriate labels for any given value. The mass assignment m_x then corresponds to the distribution of these label sets across the population. More formally, given a specific value $x \in \Omega$ an individual I identifies a subset of LA, denoted $\mathcal{D}_x^I$ to stand for the description of x given by I, as the set of words with which it is appropriate to label x. If we then allow I to vary across a population of individuals V then we naturally obtain a random set $\mathcal{D}_x$ from V into the power set of LA where $\mathcal{D}_x(I) = \mathcal{D}_x^I$. The mass assignment on labels m_x would then correspond to:

$$\forall T \subseteq LA \; m_x(T) = \frac{|\{I \in V : \mathcal{D}_x^I = T\}|}{|V|}$$

or more generally

$$m_x(T) = P\left(\{I \in V : \mathcal{D}_x^I = T\}\right)$$

where P is a probability measure on the power set of V

Similarly, the appropriateness of L to x would then correspond to the probability of encountering an individual I that deemed L to be appropriate to describe x (i.e. included L in $\mathcal{D}_x(I)$). That is:

$$\forall L \in LA \; \mu_L(x) = P\left(\{I \in V : L \in \mathcal{D}_x^I\}\right)$$

Notice that this is clearly a form of random set semantics, however, it differs from the random set interpretation of membership function proposed by Goodman and Nguyen [36], [37], [38], [72] and [73] and discussed in chapter 2 in the following sense: Using the notation introduced in chapter 2 we have that for label semantics $U_2 = LA$ whereas for the semantics of Goodman and Nguyen $U_2 = \Omega$. This simple difference has a major impact on the emergent properties of the calculus as will become apparent in the following section. It also makes the treatment of continuous systems much more straightforward, since while the universe Ω will be uncountably infinite in such cases, LA will tend to remain finite.

EXAMPLE 28 *Suppose the variable x with universe $\{1, 2, 3, 4, 5, 6\}$ gives the outcome of a single throw of a particular dice. Let LA =*

{low, medium, high} and $V = \{I_1, I_2, I_3\}$ then a possible definition of $\mathcal{D}_x$ is as follows:

$$\mathcal{D}_1^{I_1} = \mathcal{D}_1^{I_2} = \mathcal{D}_1^{I_3} = \{low\}, \mathcal{D}_2^{I_1} = \{low, medium\}, \mathcal{D}_2^{I_2} = \{low\},$$
$$\mathcal{D}_2^{I_3} = \{low\}$$
$$\mathcal{D}_3^{I_1} = \{medium\}, \mathcal{D}_3^{I_2} = \{medium\}, \mathcal{D}_3^{I_3} = \{medium, low\}$$
$$\mathcal{D}_4^{I_1} = \{medium, high\}, \mathcal{D}_4^{I_2} = \{medium\}, \mathcal{D}_4^{I_2} = \{medium\}$$
$$\mathcal{D}_5^{I_1} = \{high\}, \mathcal{D}_5^{I_2} = \{medium, high\},$$
$$\mathcal{D}_5^{I_3} = \{high\}, \mathcal{D}_6^{I_1} = \mathcal{D}_6^{I_2} = \mathcal{D}_6^{I_3} = \{high\}$$

The value of the appropriateness measure will depend on the underlying distribution on $V = \{I_1, I_2, I_3\}$, perhaps representing the weight of importance associated with the views of each individual. For instance, if we assume a uniform distribution on V then the degree of appropriateness of low as a label for 3 is given by:

$$\frac{|\{I \in V | low \in \mathcal{D}_3^I\}|}{|V|} = \frac{|\{I_3\}|}{|V|} = \frac{1}{3}$$

Overall the appropriateness degrees for each word are given by:

$$\mu_{low}(1) = \mu_{low}(2) = 1, \mu_{low}(3) = \frac{1}{3}$$

$$\mu_{medium}(2) = \frac{1}{3}, \mu_{medium}(3) = 1, \mu_{medium}(4) = 1, \mu_{medium}(5) = \frac{1}{3}$$

$$\mu_{high}(4) = \frac{1}{3}, \mu_{high}(5) = 1, \mu_{high}(6) = 1$$

Similarly, assuming a uniform prior on V we can determine mass assignments on $\mathcal{D}_x$ for $x = 1 \ldots 6$. For example, if $x = 2$ we have

$$m_2(\{low, medium\}) = \frac{|\{I \in V | \mathcal{D}_2^I = \{low, medium\}\}|}{|V|} = \frac{|\{I_1\}|}{|V|} = \frac{1}{3}$$

The mass assignments for each value of x are given by

$$m_1 = \{low\} : 1, m_2 = \{low, medium\} : \frac{1}{3}, \{low\} : \frac{2}{3}$$

$$m_3 = \{medium\} : \frac{2}{3}, \{low, medium\} : \frac{1}{3}$$

$$m_4 = \{medium, high\} : \frac{1}{3}, \{medium\} : \frac{2}{3}$$

$$m_5 = \{high\} : \frac{2}{3}, \{medium, high\} : \frac{1}{3}, m_6 = \{high\} : 1$$

3.5 Properties of Appropriateness Measures

The following results illustrate the clear relationship between λ-sets and the logical structure of the expressions that identify them. Initially, however, we introduce some basic notation. Let Val denote the set of valuations (i.e. allocations of truth values) on $\{L_1, \ldots, L_n\}$. For $v \in Val$ $v(L_i) = 1$ can be taken as meaning that L_i is an appropriate label in the current context. Let $LE^0 = \{L_1, \ldots, L_n\}$ and $LE^{n+1} = LE^n \cup \{\neg\theta, \theta \wedge \varphi, \theta \vee \varphi, \theta \rightarrow \varphi | \theta, \varphi \in LE^n\}$. Clearly we have that $LE = \bigcup_n LE^n$ and also, from a valuation v on LE^0 the truth-value, $v(\theta)$, for $\theta \in LE$ can be determined recursively in the usual way by application of the truth tables for the connectives.

DEFINITION 29 *Let* $\tau : Val \longrightarrow 2^{LA}$ *such that* $\forall v \in Val$ $\tau(v) = \{L_i | v(L_i) = 1\}$
Notice that τ *is clearly a bijection. Also note that for* $v \in Val$ $\tau(v)$ *can be associated with a Herbrand interpretation of the language* LE *(see [70]).*

LEMMA 30 $\forall \theta \in LE$ $\{\tau(v) | v \in Val, v(\theta) = 1\} = \lambda(\theta)$
Proof
We prove this by induction on the complexity of θ
Suppose $\theta \in LE^0$, *so that* $\theta = L_i$ *for some* $i \in \{1, \ldots, n\}$. *Now as* v *ranges across all valuations for which* L_i *is true, then* $\tau(v)$ *ranges across all subsets of* LA *that contain* L_i. *Hence,* $\{\tau(v) | v \in Val, v(L_i) = 1\} = \{T \subseteq LA | \{L_i\} \subseteq T\} = \lambda(L_i)$ *as required.*
Now suppose we have $\forall \theta \in LE^n$, $\{\tau(v) | v \in Val, v(\theta) = 1\} = \lambda(\theta)$ *and consider an expression* $\theta \in LE^{n+1}$ *then either* $\theta \in LE^n$ *in which case the result follows trivially or one of the following hold:*

(i) $\theta = \phi \wedge \varphi$ *where* $\phi, \varphi \in LE^n$.
  In this case $\{v \in Val | v(\phi \wedge \varphi) = 1\}$
  $= \{v \in Val | v(\phi) = 1\} \cap \{v \in Val | v(\varphi) = 1\}$.
  Therefore, $\{\tau(v) | v \in Val, v(\phi \wedge \varphi) = 1\}$
  $= \{\tau(v) | v \in Val, v(\phi) = 1\} \cap \{\tau(v) | v \in Val, v(\varphi) = 1\}$
  $= \lambda(\phi) \cap \lambda(\varphi)$ *(by the inductive hypothesis)* $= \lambda(\phi \wedge \varphi)$ *by definition 25.*

(ii) $\theta = \phi \vee \varphi$ *where* $\phi, \varphi \in LE^n$.
  In this case $\{v \in Val | v(\phi \vee \varphi) = 1\}$
  $= \{v \in Val | v(\phi) = 1\} \cup \{v \in Val | v(\varphi) = 1\}$.
  Therefore, $\{\tau(v) | v \in Val, v(\phi \vee \varphi) = 1\}$
  $= \{\tau(v) | v \in Val, v(\phi) = 1\} \cup \{\tau(v) | v \in Val, v(\varphi) = 1\}$
  $= \lambda(\phi) \cup \lambda(\varphi)$ *(by the inductive hypothesis)* $= \lambda(\phi \vee \varphi)$ *by definition 25*

(iii) $\theta = \phi \rightarrow \varphi$ *where* $\phi, \varphi \in LE^n$.
  In this case $\{v \in Val | v(\phi \rightarrow \varphi) = 1\}$
  $= \{v \in Val | v(\phi) = 0\} \cup \{v \in Val | v(\varphi) = 1\}$

$= \{v \in Val | v(\phi) = 1\}^c \cup \{v \in Val | v(\varphi) = 1\}.$
Therefore, $\{\tau(v) | v \in Val, v(\phi \to \varphi) = 1\}$
$= \{\tau(v) | v \in Val, v(\phi) = 1\}^c \cup \{\tau(v) | v \in Val, v(\varphi) = 1\}$
$= \lambda(\phi)^c \cup \lambda(\varphi)$ *(by the inductive hypothesis)* $= \lambda(\phi \to \varphi)$ *by definition 25*

(iv) $\theta = \neg\phi$ *where* $\phi \in LE^n$.
In this case $\{\tau(v) | v \in Val, v(\neg\phi) = 1\} = \{\tau(v) | v \in Val, v(\phi) = 1\}^c$
$= \lambda(\phi)^c$ *(by the inductive hypothesis)* $= \lambda(\neg\phi)$ *by definition 25.* □

THEOREM 31 *For* $\theta, \varphi \in LE$ $\theta \models \varphi$ *iff* $\lambda(\theta) \subseteq \lambda(\varphi)$
Proof
$(\Rightarrow)$ $\theta \models \varphi \Rightarrow$
$\{v \in Val | v(\theta) = 1\} \subseteq \{v \in Val | v(\varphi) = 1\} \Rightarrow$
$\{\tau(v) | v \in Val, v(\theta) = 1\} \subseteq \{\tau(v) | v \in Val, v(\varphi) = 1\} \Rightarrow$
$\lambda(\theta) \subseteq \lambda(\varphi)$ *by lemma 30.*
$(\Leftarrow)$ *Suppose* $\lambda(\theta) \subseteq \lambda(\varphi)$. *Then* $\lambda(\theta) = \{\tau(v) | v \in Val, v(\theta) = 1\}$ *and*
$\lambda(\varphi) = \{\tau(v) | v \in Val, v(\varphi) = 1\}$ *by lemma 30.*
Therefore $\{\tau(v) | v \in Val, v(\theta) = 1\} \subseteq \{\tau(v) | v \in Val, v(\varphi) = 1\} \Rightarrow$
$\{v \in Val | v(\theta) = 1\} \subseteq \{v \in Val | v(\varphi) = 1\}$ *since* τ *is a bijection.* □

A trivial corollary of theorem 31 is:

COROLLARY 32 *For* $\theta, \varphi \in LE$ $\theta \equiv \varphi$ *iff* $\lambda(\theta) = \lambda(\varphi)$

THEOREM 33 *If* $\varphi \in LE$ *is inconsistent then* $\lambda(\varphi) = \emptyset$
Proof
If $\varphi \in LE$ *is inconsistent then* $\varphi \equiv \theta \wedge \neg\theta$ *so that by corollary 32*
$\lambda(\varphi) = \lambda(\theta \wedge \neg\theta) = \lambda(\theta) \cap \lambda(\neg\theta) = \lambda(\theta) \cap \lambda(\theta)^c = \emptyset$ *by definition 25.* □

THEOREM 34

$\forall \theta \in LE, \forall x \in \Omega \ \mu_{\neg\theta}(x) = 1 - \mu_\theta(x)$

Proof

$$\mu_{\neg\theta}(x) = \sum_{T \in \lambda(\neg\theta)} m_x(T) = \sum_{T \in \lambda(\theta)^c} m_x(T) = 1 - \sum_{T \in \lambda(\theta)} m_x(T) = 1 - \mu_\theta(x)$$

From theorem 31 it follows that

$\forall x \in \Omega, \forall \theta, \varphi \in LE : \theta \models \varphi \ \mu_\varphi(x) \le \mu_\theta(x)$

Also from corollary 32 we have that

$\forall x \in \Omega, \forall \theta, \varphi \in LE : \theta \equiv \varphi \ \mu_\theta(x) = \mu_\varphi(x)$

In particular, since $\forall \theta \in LE\ \theta \equiv \theta \wedge \theta$ this means that appropriateness measures satisfy idempotence. Trivially, from theorem 33 we have that appropriateness measures satisfies the law of non-contradiction so that:

$$\forall x \in \Omega\ \forall \theta \in LE\ \mu_{\theta \wedge \neg \theta}(x) = 0$$

Given that $\neg \theta \vee \theta \equiv \neg(\theta \wedge \neg \theta)$ it follows from the above equations and theorem 34 that:

$$\forall x \in \Omega\ \forall \theta \in LE\ \mu_{\neg \theta \vee \theta}(x) = 1$$

In other words, appropriateness measures satisfy the law of excluded middle.

3.6 Functional Label Semantics

In general the proposed calculus for appropriateness measures is not functional. To see this notice that in order to determine the value of $\mu_\theta(x)$ for any $\theta \in LE$ we must first evaluate the underlying mass assignment m_x. Hence, for there to be a functional relationship between $\mu_\theta(x)$ and $\mu_{L_1}(x), \ldots, \mu_{L_n}(x)$ as required by definition 8 it must be possible to completely determine m_x from the values of $\mu_{L_1}(x), \ldots, \mu_{L_n}(x)$. However, the relationship between m_x and $\mu_{L_i}(x) : i = 1, \ldots, n$ is given by the set of equations:

$$\mu_{L_i}(x) = \sum_{T : L_i \in T} m_x(T)\ : i = 1, \ldots, n$$

These constraints alone are insufficient to determine m_x from $\mu_{L_i}(x) : i = 1, \ldots, n$. In fact the above equations corresponds to those of the fixed point coverage function of the random set $\mathcal{D}_x$ and hence, in general, only restricts m_x to an infinite family of mass assignment as shown in chapter 2 example 13.

There are, however, functional versions of label semantics arising as a result of making further assumptions regarding the relationship between m_x and $\mu_{L_i}(x) : i = 1, \ldots, n$. These additional restrictions can be formalized in terms of the concept of a mass selection function defined as follows:

DEFINITION 35 *Mass Selection Function*
Let $\mathcal{M}$ be the set of all mass assignments on 2^{LA}. Then a mass selection function (msf) is a function $\Delta : [0, 1]^n \to \mathcal{M}$ such that if $\forall x \in \Omega\ \Delta(\mu_{L_1}(x), \ldots, \mu_{L_n}(x)) = m_x$ then

$$\forall x \in \Omega\ \forall L \in LA \sum_{T \subseteq LA : L \in LA} m_x(T) = \mu_L(x)$$

In effect then a mass selection function provides sufficient information on the nature of the uncertainty regarding $\mathcal{D}_x$ to uniquely determine the mass assignment m_x from the values of the appropriateness measures on LA (see figure 3.1). Furthermore, since the value of $\mu_\theta(x)$ for any expression θ can be evaluated directly from m_x, then given a mass selection function we have a functional method for determining the appropriateness measure of any label expression from the underlying appropriateness measures of the labels. Hence, we need only define $\mu_L : \Omega \to [0,1]$ for $L \in LA$ as a fuzzy definition of each label.

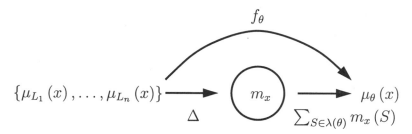

Figure 3.1: A Functional Calculus for Appropriateness Measures

We now consider two specific examples of mass selection functions and investigate the additional properties possessed by appropriateness measures in the case that these msf are adopted.

DEFINITION 36 *Consonant Mass Selection Function*
Given appropriateness measures $\mu_{L_1}(x), \ldots, \mu_{L_n}(x)$ ordered such that $\mu_{L_i}(x) \geq \mu_{L_{i+1}}(x)$ for $i = 1, \ldots, n-1$ then the consonant mass selection function identifies the mass assignment,

$$m_x(\{L_1, \ldots, L_n\}) = \mu_{L_n}(x)$$
$$m_x(\{L_1, \ldots, L_i\}) = \mu_{L_i}(x) - \mu_{L_{i+1}}(x) : i = 1, \ldots, n-1$$
and $m_x(\emptyset) = 1 - \mu_{L_1}(x)$

In terms of additional restrictions on labels, the consonant msf corresponds to the assumption that for each instance x You can identify a total ordering on the appropriateness of labels. You then evaluate Your belief values m_x about which labels are appropriate in such a way that they are consistent with this ordering. More specifically, let $\preceq_x$ denote the appropriateness ordering that You identify on LA for element x so that $L_1 \preceq_x L_2$ means that L_1 is no more appropriate than L_2 as a label for x (alternatively that L_2 is at least as appropriate as L_1). When evaluating m_x for the set of labels T You make the assumption that $m_x(T)$ is non-zero only if for every label $L_i \in T$ it holds that $L_j \in T$ for every other label L_j such that $L_i \preceq_x L_j$ (i.e. given that L_i is an appropriate label for x so must be every label L_j for which L_i is no more appropriate than L_j). Alternatively, in terms of the voting model the consonance assumption is that all individuals share a common ordering on the appropriateness of labels for a value and that the composition of $\mathcal{D}_x^I$ is consistent with this ordering for each I. More formally, supposing for each element $x \in \Omega$ the population V shares a common total ordering $\preceq_x$ as described above then in this case, when deciding on a set of appropriate labels, an individual I would be expected to be consistent with $\preceq_x$. Hence, if $L_i \in \mathcal{D}_x^I$ then L_j will also be in $\mathcal{D}_x^I$ for all labels L_j such that $L_i \preceq_x L_j$. Clearly, given such voting behaviour then as we vary individuals across V the values of $\mathcal{D}_x^I$ that actually occur will form a nested hierarchy. For instance, in the case of the dice problem described in example 28 possible appropriateness orderings for values $x = 1, \ldots, 6$ are as follows:

$$high \preceq_1 medium \preceq_1 low, high \preceq_2 medium \preceq_2 low$$

$$high \preceq_3 low \preceq_3 medium, low \preceq_4 high \preceq_4 medium$$

$$low \preceq_5 medium \preceq_5 high, low \preceq_6 medium \preceq_6 high$$

Hence, for any individual I, if I decides that low is an appropriate label for 3 ($low \in \mathcal{D}_3^I$) then to be consistent with the ordering $\preceq_3$ they must also decide that $medium$ is an appropriate label for 3 ($medium \in \mathcal{D}_3^I$) since $medium$ is at least as appropriate as low as a label for 3.

Notice, that the consonance assumption for random sets on labels is in one sense weaker than the corresponding assumption for random sets on the universe Ω (see definition 14), since the latter requires individuals to maintain the same level of specificity across all values in Ω. To see this more clearly recall example 28 and observe that m_x is consonant $\forall x \in \{1, \ldots, 6\}$. Now for each member $I \in V$ the extension (associated subset of Ω) of, say *medium* is given by $\{x \in \Omega | medium \in \mathcal{D}_x^I\}$. Hence, we obtain $\{2, 3, 4\}, \{3, 4, 5\}$ and $\{3, 4\}$ for I_1, I_2 and I_3 respectively. Clearly, however, this does not form a nested hierarchy.

THEOREM 37 $\forall x \in \Omega$ *let the mass assignment on labels m_x be determined from $\mu_L(x) : L \in LA$ according to the consonant msf. Then for $L_1, \ldots, L_k \in LA$, where $1 \leq k \leq n$, we have that*

$$\forall x \in \Omega \; \mu_{L_1 \wedge \ldots \wedge L_k}(x) = \min(\mu_{L_1}(x), \ldots, \mu_{L_k}(x))$$

Proof
By definition 25 we have that

$$\lambda(L_1 \wedge \ldots \wedge L_k) = \lambda(L_1) \cap \ldots \cap \lambda(L_k)$$
$$= \{T \subseteq LA : \{L_1\} \subseteq T\} \cap \ldots \cap \{T \subseteq LA : \{L_k\} \subseteq T\} =$$
$$\{T \subseteq LA : \{L_1, \ldots, L_k\} \subseteq T\}$$

Hence,

$$\forall x \in \Omega \; \mu_{L_1 \wedge \ldots \wedge L_k}(x) = \sum_{T \subseteq LA : \{L_1, \ldots, L_k\} \subseteq T} m_x(T)$$

For any x, since m_x is a consonant mass assignment then it must have the form $m_x = M_0 : m_0, \ldots, M_t : m_t$ where $M_i \subset M_{i+1}$ for $i = 0, \ldots, t - 1$.
Now suppose w.l.o.g that $\mu_{L_1}(x) \leq \ldots \leq \mu_{L_k}(x)$ then $\{L_1\} \subseteq M_i$ iff $\{L_1, \ldots, L_k\} \subseteq M_i$ for $i = 0, \ldots, t$.
Therefore

$$\mu_{L_1 \wedge \ldots \wedge L_k}(x) = \sum_{T : \{L_1\} \subseteq T} m_x(T) = \mu_{L_1}(x) = \min(\mu_{L_1}(x), \ldots, \mu_{L_k}(x)) \; \square$$

THEOREM 38 *$\forall x \in \Omega$ let the mass assignment on labels m_x be determined from $\mu_L(x) : L \in LA$ according to the consonant msf. Then for $L_1, \ldots, L_k \in LA$, where $1 \leq k \leq n$, we have that*

$$\forall x \in \Omega \; \mu_{L_1 \vee \ldots \vee L_k}(x) = \max(\mu_{L_1}(x), \ldots, \mu_{L_k}(x))$$

Proof *Similar to that of theorem 37.*

A summary of the properties of appropriateness measures under the consonant msf, taking into account results from both this and the previous section, is a follows:

Properties of Appropriateness Measures under the Consonant msf

- If $\theta \models \varphi$ then $\forall x \in \Omega \; \mu_\theta(x) \leq \mu_\varphi(x)$

- If $\theta \equiv \varphi$ then $\forall x \in \Omega \; \mu_\theta(x) = \mu_\varphi(x)$

- If θ is a tautology then $\forall x \in \Omega \; \mu_\theta(x) = 1$

- If θ is a contradiction then $\forall x \in \Omega \; \mu_\theta(x) = 0$

- $\forall x \in \Omega \; \mu_{\neg\theta}(x) = 1 - \mu_\theta(x)$

- Applying the consonant msf then it holds that $\forall L_1, \ldots, L_k \in LA$, $\forall x \in \Omega \; \mu_{L_1 \wedge \ldots \wedge L_k}(x) = \min(\mu_{L_1}(x), \ldots, \mu_{L_k}(x))$

- Applying the consonant msf then it holds that $\forall L_1, \ldots, L_k \in LA$, $\forall x \in \Omega \; \mu_{L_1 \vee \ldots \vee L_k}(x) = \max(\mu_{L_1}(x), \ldots, \mu_{L_k}(x))$

In the light of these properties we now re-examine the distinction between a functional calculus and a truth-functional calculus by focusing on appropriateness measures under the consonant msf. By definition of a mass selection function this calculus is clearly functional but, as we shall see, it cannot be truth-functional. This can be seen both indirectly, from Dubois and Prade's theorem [20] (chapter 2, theorem 6), or directly from the properties of the calculus. In terms of Dubois and Prade's result we note that appropriateness measures

satisfy both idempotence and the law of excluded middle, yet they are not restricted to binary values as would be the case if their underlying calculus was
truth-functional. More directly, notice that while appropriateness measures under the consonant msf satisfy Zadeh's [110] min-max rules for conjunctions and
disjunctions of labels, this is not generally the case for conjunction and disjunctions of compound expressions from LE. For example, consider $\mu_{L_i \wedge \neg L_j}(x)$.
From definition 25 we have that $\lambda(L_i \wedge \neg L_j) = \lambda(L_i) \cap \lambda(L_j)^c$ and hence

$$\mu_{L_i \wedge \neg L_j}(x) = \sum_{T: L_i \in T, L_j \notin T} m_x(T)$$

Given the consonance assumption we know that $m_x = M_0 : m_0, \ldots, M_t : m_t$
where $M_r \subset M_{r+1}$ for $r = 0, \ldots, t-1$. Now suppose that $\mu_{L_i}(x) \leq \mu_{L_j}(x)$
then for all $r = 0, \ldots, t$ if $L_i \in M_r$ then $L_j \in M_r$ and hence $\mu_{L_i \wedge \neg L_j}(x) = 0$.
Alternatively if $\mu_{L_i}(x) \geq \mu_{L_j}(x)$ then

$$\mu_{L_i \wedge \neg L_j}(x) = \sum_{T: L_i \in T, L_j \notin T} m_x(S) = \sum_{T: L_i \in T} m_x(T) - \sum_{T: L_j \in T} m_x(T) =$$

$$\mu_{L_i}(x) - \mu_{L_j}(x)$$

This can be summarised by the expression $\mu_{L_i \wedge \neg L_j}(x) = \max(0, \mu_{L_i}(x) - \mu_{L_j}(x))$ which is not in general the same as $\min(\mu_{L_i}(x), 1 - \mu_{L_j}(x))$ as would
be given by the truth-functional calculus for which $f_\wedge(a, b) = \min(a, b)$,
$f_\vee(a, b) = \max(a, b)$ and $f_\neg(a) = 1 - a$. As an aside, we note that this
result gives some insight into the behaviour of implication in label semantics,
at least at the level of individual labels. For instance, we have that $L_i \rightarrow L_j$ is
logically equivalent to $\neg(L_i \wedge \neg L_j)$ and hence $\mu_{L_i \rightarrow L_j}(x) = 1 - \mu_{L_i \wedge \neg L_j}(x) = 1 - \max(0, \mu_{L_i}(x) - \mu_{L_j}(x)) = \min(1, 1 - \mu_{L_i}(x) + \mu_{L_j}(x))$. This corresponds to Lukasiewicz implication (see [54] for an exposition) although it only
applies here at the label level and not for more complex expressions.

EXAMPLE 39 *Let* $LA = \{small(s), medium(m), large(l)\}$, $\Omega = [0, 10]$
and let μ_s, μ_m *and* μ_l *be trapezoidal functions (see figure 3.2) defined by*

$$\mu_{small}(x) = \begin{cases} 1 : x \in [0, 2] \\ 2 - \frac{x}{2} : x \in (2, 4] \\ 0 : x > 4 \end{cases}$$

$$\mu_{medium}(x) = \begin{cases} 0 : x < 2 \\ \frac{x}{2} - 1 : x \in [2, 4] \\ 1 : x \in (4, 6] \\ 4 - \frac{x}{2} : x \in (6, 8] \\ 0 : x > 8 \end{cases}$$

$$\mu_{large}(x) = \begin{cases} 0 : x < 6 \\ \frac{x}{2} - 3 : x \in [6,8] \\ 1 : x > 8 \end{cases}$$

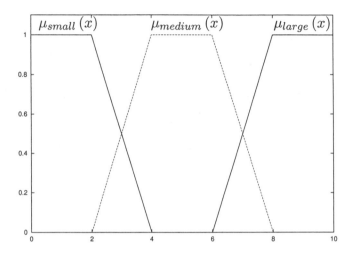

Figure 3.2: Appropriateness measures for, from left to right, *small*, *medium* and *large*

Allowing x to vary across the universe $[0,10]$ we obtain the following defi-nition of $m_x(T)$ as a function of x for each $T \subseteq LA$ (see figure 3.3). Since, the overlap between the labels small, medium and large is restricted, as can be seen from figure 3.2, only a subset of the power set of LA have non-zero mass for any x. For example, since the labels small and large do not overlap then $\forall x \in \Omega\ m_x(\{s,l\}) = 0$. This observation motivates the definition of focal elements as given in the sequel (definition 40).

$$m_x(\{small\}) = \begin{cases} 1 : x \in [0,2] \\ 3 - x : x \in (2,3] \\ 0 : x > 3 \end{cases}$$

$$m_x(\{small, medium\}) = \begin{cases} 0 : x < 2 \\ \frac{x}{2} - 1 : x \in [2,3] \\ 2 - \frac{x}{2} : x \in (3,4] \\ 0 : x > 4 \end{cases}$$

$$m_x\left(\{medium\}\right) = \begin{cases} 0 : x < 3 \\ x - 3 : x \in [3,4] \\ 1 : x \in (4,6] \\ 7 - x : x \in (6,7] \\ 0 : x > 7 \end{cases}$$

$$m_x\left(\{medium, large\}\right) = \begin{cases} 0 : x < 6 \\ \frac{x}{2} - 3 : x \in [6,7] \\ 4 - \frac{x}{2} : x \in (7,8] \\ 0 : x > 8 \end{cases}$$

$$m_x\left(\{large\}\right) = \begin{cases} 0 : x < 7 \\ x - 7 : x \in [7,8] \\ 1 : x > 8 \end{cases}$$

$$m_x\left(\emptyset\right) = \begin{cases} 0 : x < 2 \\ \frac{x}{2} - 1 : x \in [2,3] \\ 2 - \frac{x}{2} : x \in (3,4] \\ 0 : x \in (4,6] \\ \frac{x}{2} - 3 : x \in (6,7] \\ 4 - \frac{x}{2} : x \in (7,8] \\ 0 : x > 8 \end{cases}$$

Given the above values for m_x we can then evaluate $\mu_{m \wedge \neg l}(x)$ as follows:

$$\mu_{medium \wedge \neg large}(x) =$$

$$m_x\left(\{small, medium\}\right) + m_x\left(\{medium\}\right) = \begin{cases} 0 : x < 2 \\ \frac{x}{2} - 1 : x \in [2,4] \\ 1 : x \in (4,6] \\ 7 - x : x \in (6,7] \\ 0 : x > 7 \end{cases}$$

Or alternatively, since we are applying the consonant msf, then from the above discussion:

$$\mu_{medium \wedge \neg large}(x) = \max\left(0, \mu_{medium}(x) - \mu_{large}(x)\right)$$

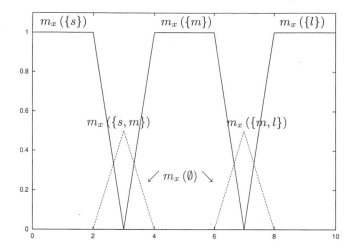

Figure 3.3: Mass assignments for varying x under the consonant msf; shown from left to right, $m_x(\{small\})$, $m_x(\{small, medium\})$, $m_x(\{medium\})$, $m_x(\{medium, large\})$ and $m_x(\{large\})$; $m_x(\emptyset)$ is equal to $m_x(\{small, medium\})$ for $x \in [2, 4]$, is equal to $m_x(\{medium, large\})$ for $x \in [6, 8]$ and is zero otherwise

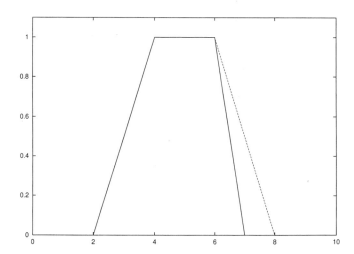

Figure 3.4: Appropriateness Measure $\mu_{medium \wedge \neg large}(x)$ under the consonant msf (solid line) and $\min(\mu_{medium}(x), 1 - \mu_{large}(x)) = \mu_{medium}(x)$ (dashed line) corresponding to $\mu_{medium \wedge \neg large}(x)$ in truth-functional fuzzy logic

Figure 3.4 clearly illustrates the difference between label semantics and fuzzy logic when evaluating compound expressions such as, in this case, medium ∧ ¬large. It is interesting to note that using truth-functional fuzzy logic based on min as the conjunction function, the two statements 'x is medium' and 'x *is medium but not large' provide exactly the same information (i.e. they have the same memberships). In other words, the extra information that 'x is not large' tells us nothing. This seems highly counter intuitive. On the other hand, in label semantics $\mu_{medium \wedge \neg large}(x)$ is zero for all values greater than seven since for such values the only sets of appropriate labels with non-zero mass containing medium also contain large.*

DEFINITION 40 *Set of Focal Elements*
Given labels LA together with associated mass assignments $m_x \; \forall x \in \Omega$, the set of focal elements for LA is given by

$$\mathcal{F} = \{T \subseteq LA : \exists x \in \Omega, \; m_x(T) > 0\}$$

For instance, in example 39

$$\mathcal{F} = \{\{s\}, \; \{s,m\}, \; \{m\}, \; \{m,l\}, \; \{l\}, \; \emptyset\}$$

In addition to the consonant msf we also consider in detail the following msf based on an underlying assumption of independence between labels.

DEFINITION 41 *Independent Mass Selection Function*
Given appropriateness measures $\mu_{L_1}(x), \ldots, \mu_{L_n}(x)$ then the independent mass selection function identifies the mass assignment:

$$\forall T \subseteq LA \; m_x(T) = \prod_{L \in T} \mu_L(x) \times \prod_{L \notin T} (1 - \mu_L(x))$$

The independent msf simply assumes that when judging the appropriateness of a label You do not take into account the level of appropriateness of any other label. Although this seems extreme, it may be reasonable in the case that labels relate to different facets of the instance. For example, the appropriateness of the label 'thin' might be assumed to be independent of the appropriateness of the label 'rich'.

THEOREM 42 *If $\forall T \subseteq LA \; m_x(T) = \prod_{L \in T} \mu_L(x) \times \prod_{L \notin T}(1 - \mu_L(x))$ then $\forall L_i, \ldots, L_k \in LA : \mu_{L_1 \wedge \ldots \wedge L_k}(x) = \prod_{i=1}^{n} \mu_{L_i}(x)$*

Proof

$\forall L_i, L_j \in LA : j \neq i$

$$\mu_{L_1 \wedge \ldots \wedge L_j}(x) = \sum_{T:L_1 \in T, \ldots, L_k \in T} m_x(T) =$$

$$\sum_{T:L_1 \in T, \ldots, L_k \in T} \left(\prod_{L \in T} \mu_L(x) \times \prod_{L \notin T} (1 - \mu_L(x)) \right)$$

$$= \sum_{T:L_1 \in T, \ldots, L_k \in T} \prod_{i=1}^{k} \mu_{L_i}(x) \left(\prod_{L \in T: L \neq L_1, \ldots, L \neq L_k} \mu_L(x) \times \prod_{L \notin T} (1 - \mu_L(x)) \right)$$

$$= \prod_{i=1}^{k} \mu_{L_i}(x) \sum_{T:L_1 \in S, \ldots, L_k \in S} \left(\prod_{L \in T: L \neq L_1, \ldots L \neq L_k} \mu_L(x) \times \prod_{L \notin T} (1 - \mu_L(x)) \right)$$

$$= \prod_{i=1}^{k} \mu_{L_i}(x) \sum_{T \subseteq LA - \{L_1, \ldots L_k\}} \left(\prod_{L \in T} \mu_L(x) \times \prod_{L \notin T} (1 - \mu_L(x)) \right) = \prod_{i=1}^{k} \mu_{L_i}(x)$$

Since $\forall L \in LA \ \mu_L(x) \in [0,1]$ *implies that*

$$\sum_{S \subseteq LA - \{L_1, \ldots, L_k\}} \left(\prod_{L \in T} \mu_L(x) \times \prod_{L \notin T} (1 - \mu_L(x)) \right) = 1 \ \square$$

THEOREM 43 *If* $\forall T \subseteq LA \ m_x(T) = \prod_{L \in T} \mu_L(x) \times \prod_{L \notin T} (1 - \mu_L(x))$ *then* $\forall L_i, \ldots, L_k \in LA$:

$\mu_{L_1 \vee \ldots \vee L_k}(x) = \sum_{T \subseteq \{L_1, \ldots, L_k\}} (-1)^{|T|-1} \prod_{L \in T} \mu_L(x)$

Proof *Similar to that of theorem 42.*

A summary of the properties of appropriateness measures under the independent msf, taking into account results from both this and the previous section, is a follows:

Properties of Appropriateness Measures under the Independent msf

- If $\theta \models \varphi$ then $\forall x \in \Omega \ \mu_\theta(x) \leq \mu_\varphi(x)$

- If $\theta \equiv \varphi$ then $\forall x \in \Omega \ \mu_\theta(x) = \mu_\varphi(x)$

- If θ is a tautology then $\forall x \in \Omega \ \mu_\theta(x) = 1$

- If θ is a contradiction then $\forall x \in \Omega \ \mu_\theta(x) = 0$

- $\forall x \in \Omega \ \mu_{\neg\theta}(x) = 1 - \mu_\theta(x)$

- Applying the independent msf then it holds that $\forall L_1, \ldots, L_k \in LA$, $\forall x \in \Omega \ \mu_{L_1 \wedge \ldots \wedge L_k}(x) = \prod_{i=1}^{n} \mu_{L_i}(x)$

- Applying the independent msf then it holds that $\forall L_1, \ldots, L_k \in LA$, $\forall x \in \Omega \ \mu_{L_1 \vee \ldots \vee L_k}(x) = \sum_{T \subseteq \{L_1, \ldots, L_k\}} (-1)^{|T|-1} \mu_{\bigwedge_{L \in T} L}(x)$

We can again use Dubois and Prade's [20] theorem to see that this calculus is also functional but not truth-functional in the same way as under the consonant msf. However, there are some differences between these two calculi. For example, unlike for the consonant msf $\mu_{L_i \wedge \neg L_j}(x) = \mu_{L_i}(x) \times \left(1 - \mu_{L_j}\right)$ which is in agreement with the truth-functional calculus with $f_\neg(a) = 1 - a$ and $f_\wedge(a, b) = a \times b$. Interestingly, this means that $\mu_{L_i \to L_j} = 1 - \mu_{L_i}(x) + \mu_{L_i}(x)\mu_{L_j}(x)$ which corresponds to the Reichenbach implication operator (see [54] for an exposition), although, once again, it applies only at the label level and not for more general compound expressions. For a direct failure of truth-functionality we need only consider the idempotence properties of the calculus. In particular, for appropriateness measures under the independent msf $\mu_{L \wedge L}(x) = \mu_L(x)$ and not $\mu_L(x)^2$ as under the truth-functional calculus with $f_\wedge(a, b) = a \times b$.

EXAMPLE 44 *Given LA and appropriateness measures for small, medium and large as defined in example 39 then by applying the independent mass selection function we obtain the following mass assignment m_x for varying x (see figure 3.5):*

$$m_x(\{small\}) = \begin{cases} 1 : x \in [0, 2] \\ \frac{(x-4)^2}{4} : x \in (2, 4] \\ 0 : x > 4 \end{cases}$$

$$m_x(\{small, medium\}) = \begin{cases} 0 : x < 2 \\ -\frac{(x-4)(x-2)}{4} : x \in [2, 4] \\ 0 : x > 4 \end{cases}$$

$$m_x(\{medium\}) = \begin{cases} 0 : x < 2 \\ \frac{(x-4)^2}{4} : x \in [2, 4] \\ 1 : x \in (4, 6] \\ \frac{(x-8)^2}{4} : x \in (6, 8] \\ 0 : x > 8 \end{cases}$$

$$m_x \left(\{medium, large\} \right) = \begin{cases} 0 : x < 6 \\ -\frac{(x-8)(x-6)}{4} : x \in [6,8] \\ 0 : x > 8 \end{cases}$$

$$m_x \left(\{large\} \right) = \begin{cases} 0 : x < 6 \\ \frac{(x-6)^2}{4} : x \in [6,8] \\ 1 : x > 8 \end{cases}$$

$$m_x \left(\emptyset \right) = \begin{cases} 0 : x < 2 \\ -\frac{(x-4)(x-2)}{4} : x \in [2,4] \\ 0 : x \in (4,6] \\ -\frac{(x-8)(x-6)}{4} : x \in (6,8] \\ 0 : x > 8 \end{cases}$$

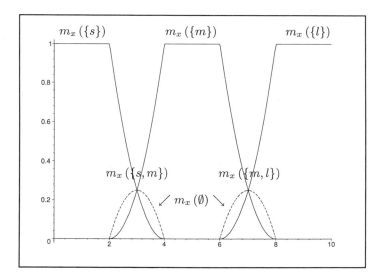

Figure 3.5: Mass assignments for varying x under the independent msf; shown from left to right, $m_x(\{small\})$, $m_x(\{small, medium\})$, $m_x(\{medium\})$, $m_x(\{medium, large\})$ and $m_x(\{large\})$; $m_x(\emptyset)$ is equal to $m_x(\{small, medium\})$ for $x \in [2,4]$, is equal to $m_x(\{medium, large\})$ for $x \in [6,8]$ and is zero otherwise

The appropriateness measures for compound expressions in LE can then be evaluated directly from these mass assignments. For example, the appropriateness measure for medium $\land \neg large$ (see figure 3.6) is given by:

$$\mu_{medium \wedge \neg large}(x) =$$

$$m_x(\{small, medium\}) + m_x(\{medium\}) = \begin{cases} 0 : x < 2 \\ \frac{x}{2} - 1 : x \in [2, 4] \\ 1 : x \in (4, 6] \\ \frac{(x-8)^2}{4} : x \in (6, 8] \\ 0 : x > 8 \end{cases}$$

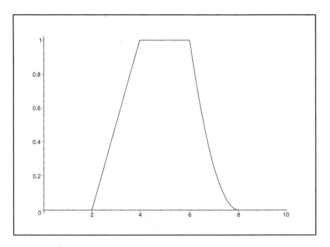

Figure 3.6: Appropriateness Measure $\mu_{medium \wedge \neg large}(x)$ under the independent msf.

3.7 Relating Appropriateness Measures to Dempster-Shafer Theory

Fuzzy sets have long been linked to Dempster-Shafer theory, originally through the work on Kampe de Feriet [51] who notes that when membership function $\chi_\theta(x)$ is interpreted as the single point coverage function of a random set with mass assignment m then:

$$\chi_\theta(x) = \sum_{S \subseteq \Omega : S \subseteq \{x\}} m(S) = Bel(\{x\}) = \sum_{S \subseteq \Omega : S \cap \{x\} \neq \emptyset} m(S) = Pl(\{x\})$$

There are also a number of clear connections between appropriateness measures and Shafer-Dempster theory [95]. Suppose we simply view m_x as a conditional mass assignment on 2^{LA} given value x. In this case, for any labels $L_1, \ldots, L_k \in LA$ the appropriateness of the disjunction $L_1 \vee \cdots \vee L_k$ as a description of x is given by:

$$\mu_{L_1 \vee \cdots \vee L_k}(x) = \sum_{T : \{L_1, \ldots, L_k\} \cap T \neq \emptyset} m_x(T) = Pl(\{L_1, \ldots, L_k\} | x)$$

Similarly the appropriateness of the conjunction $L_1 \wedge \cdots \wedge L_k$ as a description of x is given by:

$$\mu_{L_1 \wedge \cdots \wedge L_k}(x) = \sum_{T:\{L_1,\ldots,L_k\} \subseteq T} m_x(T) = Q(\{L_1,\ldots,L_k\}|x)$$

Here Q denotes the commonality function for m_x where for subset T, $Q(T)$ represents the total mass that can be moved freely to every element of T [95]. Finally, for the case of negation we have that appropriateness of $\neg(L_1 \vee \cdots \vee L_k)$ as a label for x is given by:

$$\mu_{\neg(L_1 \vee \cdots \vee L_k)}(x) = \mu_{\neg L_1 \wedge \cdots \wedge \neg L_k}(x) = \sum_{T:\{L_1,\ldots,L_k\} \cap T = \emptyset} m_x(T)$$

$$= \sum_{T:T \subseteq \{L_1,\ldots,L_k\}^c} m_x(T) = Bel(\{L_1,\ldots,L_k\}^c|x)$$

The above together with the consonance assumption means that $\mu_{L_1 \vee \cdots \vee L_k}(x)$ can be interpreted as $\Pi(\{L_1,\ldots,L_k\}|x)$ where Π is a possibility measure on 2^{LA}.

We should not make the mistake, however, in thinking that appropriateness measures are somehow special cases of D-S belief (plausibility or commonality) functions as is the case with possibility measures. The general method of evaluating appropriateness measures by summing over λ-sets has no equivalent in D-S theory. For example,

$$\mu_{L_i \wedge \neg L_j}(x) = \sum_{T:T \in \lambda(L_i \wedge \neg L_j)} m_x(T) = \sum_{T:L_i \in T, L_j \notin T} m_x(T)$$

does not correspond to either a belief, plausibility or commonality function from D-S theory. It is instead the case that $\mu_\theta(x)$ can be interpreted as particular measures from D-S theory for a certain restricted subset of expressions $\theta \in LE$. None the less, this limited form of relationship is useful in the development of label semantics as we shall see in the following section.

3.8   Mass Selection Functions based on t-norms

We have already seen that the calculus for appropriateness measures can be consistent with the restricted use of t-norms and t-conorms at the label level. For instance, for the consonant msf the appropriateness measures of conjunction and disjunction of labels correspond respectively to the min and max of the appropriateness measures of the relevant labels. Similarly, the independent msf is consistent with the use of the product t-norm and its associated dual t-conorm at the label level. The existence of such mass selection functions naturally raises the question as to whether there exists a msf consistent with any t-norm/t-conorm pair in this manner? Or more formally, can we always find a mass selection function such that for any t-norm $f_\wedge$ and dual t-conorm $f_\vee$ the associated calculus for appropriateness measures will satisfy the following equations?

$$\forall L_1, \ldots, L_k \in LA \; \mu_{L_1 \wedge \ldots \wedge L_k}(x) = f_\wedge(\mu_{L_1}(x), \ldots, \mu_{L_k}(x))$$
$$\forall L_1, \ldots, L_k \in LA \; \mu_{L_1 \vee \ldots \vee L_k}(x) = f_\vee(\mu_{L_1}(x), \ldots, \mu_{L_k}(x))$$

Now we can immediately place a rather strong restrictions on the type of t-norm/t-conorm pairs that can be consistent with appropriateness measures in this way. To see this note that, trivially from definitions 25 and 27 it holds that:

$$\forall L_i, L_j \in LA \; \mu_{L_i \vee L_j}(x) = \mu_{L_i}(x) + \mu_{L_j}(x) - \mu_{L_i \wedge L_j}(x)$$

Since, this must hold for any values of $\mu_{L_i}(x)$ and $\mu_{L_j}(x)$, hence for the calculus to be consistent with $f_\wedge$ and $f_\vee$ at the label level as above it follows that:

$$\forall y_1, y_2 \in [0, 1] \; f_\vee(y_1, y_2) = y_1 + y_2 - f_\wedge(y_1, y_2)$$

This is Frank's equation [31] and given that $f_\vee$ is the dual t-conorm of $f_\wedge$ then $f_\wedge$ is restricted to ordinal sums of the family of Frank's t-norms as described in chapter 2, definition 5. In fact for simplicity we shall now consider only Frank's t-norms, an assumption for which the results of [31] provides at least partial justification. Given the relationship between appropriateness measures and Dempster-Shafer theory as outlined in the previous section we can use a result due to Shafer [95] to find a mapping between such Frank's t-norms and the underlying mass assignment on labels m_x. In [95] Shafer showed that there is a direct relationship between a mass assignment m on 2^Ω and its associated commonality function Q captured by the following inversion formula:

$$\forall S \subseteq \Omega \; m(S) = \sum_{R \subseteq \Omega : R \supseteq S} (-1)^{|R-S|} Q(R)$$

Now from the previous section we know that appropriateness measures of conjunctions of labels are a special case of commonality measures in the sense that:

$$\mu_{L_1 \wedge \ldots \wedge L_k}(x) = Q(\{L_1, \ldots, L_k\} | x)$$

Hence, we can apply Shafer's inversion formula [95] so that:

$$\forall T \subseteq LA \; m_x(T) = \sum_{R \subseteq LA : R \supseteq T} (-1)^{|R-T|} \mu_{\wedge_{L \in R} L}(x)$$
$$= \sum_{R \subseteq LA : R \supseteq T} (-1)^{|R-T|} f_\wedge(\mu_L(x) : L \in R)$$

Interestingly the above equation would seem to define a particular family of mass selection functions that can be generated from Frank's t-norms. There is, however, an additional complication in that not all Frank's t-norms can be used to generate consistent msf in this way. To see this notice that in order for the msf generated by $f_\wedge$ to be consistent it must hold that:

$$\forall T \subseteq LA \; m_x(T) \geq 0$$

and hence $\forall T \subseteq LA \displaystyle\sum_{R \subseteq LA : R \supseteq T} (-1)^{|R-T|} f_\wedge(\mu_L(x) : L \in R) \geq 0$

Now consider the following counter example for the Lukasiewicz t-norm $f_\wedge(a,b) = \max(0, a+b-1)$ corresponding to the Frank t-norm obtaining by letting the parameter s tend to ∞.

EXAMPLE 45 *Let* $LA = \{L_1, L_2, L_3\}$ *and* $x \in \Omega$ *such that* $\mu_{L_1}(x) = 0.2$, $\mu_{L_2}(x) = 0.7$ *and* $\mu_{L_3}(x) = 0.6$. *Now*
$\wedge_t(\{\mu_{L_1}(x), \mu_{L_2}(x), \mu_{L_3}(x)\}) = \max(0, 0.2 + 0.6 + 0.7 - 2) = 0$,
$\wedge_t(\{\mu_{L_1}(x), \mu_{L_2}(x)\}) = \max(0, 0.2 + 0.7 - 1) = 0$,
$\wedge_t(\{\mu_{L_1}(x), \mu_{L_3}(x)\}) = \max(0, 0.2 + 0.6 - 1) = 0$
and $\wedge_t(\{\mu_{L_2}(x), \mu_{L_3}(x)\}) = \max(0, 0.7 + 0.6 - 1) = 0.3$
Hence, applying the inversion formula we obtain
$m_x(\emptyset) = 1 - \mu_{L_1}(x) - \mu_{L_2}(x) - \mu_{L_3}(x) + \wedge_t(\{\mu_{L_1}(x), \mu_{L_2}(x)\}) +$
$\wedge_t(\{\mu_{L_1}(x), \mu_{L_3}(x)\}) + \wedge_t(\{\mu_{L_2}(x), \mu_{L_3}(x)\}) -$
$\wedge_t(\{\mu_{L_1}(x), \mu_{L_2}(x), \mu_{L_3}(x)\}) = 1 - 0.2 - 0.6 - 0.7 + 0.3 = -0.2$

Hence, there is no mass selection function for which the associated calculus for appropriateness measures is consistent with the Lukasiewicz t-norm and dual t-conorm at the label level. However, there do exist mass selection functions which are consistent with the Lukasiewicz t-norm for particular ranges of numerical values of $\mu_{L_1}(x), \ldots, \mu_{L_n}(x)$ as is shown by the following theorems:

THEOREM 46 *For* $x \in \Omega$ *such that* $\sum_{i=1}^n \mu_{L_i}(x) \leq 1$ *then* $\forall L_1, \ldots, L_k \in$ $LA \; \mu_{L_i \wedge \ldots \wedge L_k}(x) = \max(0, \sum_{i=1}^k \mu_{L_i}(x) - (k-1))$ *iff the underlying mass assignment on labels* m_x *has the following form:* $\{L_i\} : \mu_{L_i}(x)$ *for* $i = 1, \ldots, n$ *and* $\emptyset : 1 - \sum_{i=1}^n \mu_{L_i}(x)$.

Proof

($\Leftarrow$)

$$\forall L_1, \ldots, L_k \in LA \; \mu_{L_1 \wedge \ldots \wedge L_k}(x) = \sum_{T : L_1 \in T, \ldots, L_k \in T} m_x(T) = 0$$

since there are no sets with non-zero mass containing more than one element. Also $\max(0, \mu_{\sum_{i=1}^k L_i}(x) - (k-1)) = 0$ *because* $\sum_{i=1}^n \mu_{L_i}(x) \leq$

$1 \Rightarrow \sum_{i=1}^{k} \mu_{L_i}(x) \leq 1$

In addition, we have that $\forall L_i \in LA \sum_{T:L_i \in T} m_x(T) = m_x(\{L_i\}) = \mu_{L_i}(x)$
as required.

$(\Rightarrow)$
For any pair of labels $L_1, L_2 \in LA$

$$\forall L_1, L_2 \in LA \sum_{T:L_1 \in T, L_2 \in T} m_x(T) = \max(0, \mu_{L_1}(x) + \mu_{L_2}(x) - 1) = 0$$

by the above argument. This means that $\forall T \subseteq LA : |T| > 1 \; m_x(T) = 0$ *and hence that the only possible sets with non-zero mass are* $\{L_i\} : i = 1, \ldots, n$ *and* $\emptyset$. *From this and the fact that* $\forall L_i \in LA \sum_{T:L_i \in T} m_x(T) = \mu_{L_i}(x)$ *we can infer that* $m_x(\{L_i\}) = \mu_{L_i}(x)$ *for* $i = 1, \ldots, n$ *and then* $m_x(\emptyset) = 1 - \sum_{i=1}^{n} \mu_{L_i}(x)$ *follows from the fact that* $\sum_{T \subseteq LA} m_x(T) = 1$. $\square$

Any msf partially defined as in theorem 46 on this particular region of the appropriateness measure space is characterised by very conservative behaviour where You are at most willing to agree only to a single member of LA being appropriate as a label for x. In the limit case where $\sum_{i=1}^{n} \mu_{L_k}(x) = 1$ then m_x as selected by this msf corresponds to a probability distribution on LA.

THEOREM 47 *For* $x \in \Omega$ *such that* $\sum_{i=1}^{n} \mu_{\neg L_i}(x) \leq 1$ *then* $\forall L_1, \ldots, L_k \in LA \; \mu_{L_1 \wedge \ldots \wedge L_k}(x) = \max(0, \sum_{i=1}^{k} \mu_{L_i}(x) - (k-1))$ *iff the underlying mass assignment on labels* m_x *has the following form :* $\{L_1, \ldots, L_n\} - \{L_i\} : 1 - \mu_{L_i}(x)$ *for* $i = 1, \ldots, n$ *and* $\{L_1, \ldots, L_n\} : \sum_{i=1}^{n} \mu_{L_i}(x) + 1 - n$.

The msf partially defined as in theorem 47 on this second region of the input space is characterised by somewhat indecisive behaviour where You are only willing to eliminate at most one element of LA from the set of appropriate labels for x.

Since we have shown that there is no mass selection function for which the calculus of appropriateness measures is universally consistent with the Lukasiewicz t-norm/t-conorm pair it is natural to consider which members of the Frank's family of t-norms do generate consistent msf according to the inversion formula described previously. In fact this remains an open question although the following theorems do shed some light on the matter [65].

THEOREM 48 *There is no consistent msf that can be generated according to the inversion formula from a Frank's t-norm with* $s \geq 2$
Proof
Let $LA = \{L_1, L_2, L_3\}$ *and let* μ *be the appropriateness measure generated by* $\wedge_s$ *for* $s > 1$ *such that for some* $x \in \Omega \; \mu_{L_1}(x) = \mu_{L_2}(x) = \mu_{L_3}(x) = y$ *for*

some $y \in [0, 1]$. Then from the inversion formula it follows that

$$m_x(\emptyset) = 1 - 3y + 3\wedge_s(y, y) - \wedge_s(y, y, y) = 1 - 3y + \log_s \left(\frac{\left[1 + \frac{(s^y - 1)^2}{s - 1}\right]^3}{1 + \frac{(s^y - 1)^3}{(s - 1)^2}} \right)$$

We now show that for $s \geq 2$ there exists a value of y for which

$$1 - 3y + \log_s \left(\frac{\left[1 + \frac{(s^y - 1)^2}{s - 1}\right]^3}{1 + \frac{(s^y - 1)^3}{(s - 1)^2}} \right) < 0$$

Putting $z = s^y - 1$ and $w = s - 1$ this corresponds to:

$$\frac{(w + z^2)^3}{w(w^2 + z^3)} < \frac{(z + 1)^3}{w + 1} \text{ where } w \geq 0 \text{ and } z \in [0, w]$$

$\Rightarrow (w + z^2)(w + 1) < w(w^2 + z^3)(x + 1)^3$
$\Rightarrow z^6 - 3wz^5 + 3w^2z^4 - (w^3 + w)z^3 + 3w^2z^2 - 3w^3z + z^4 < 0$
$\Rightarrow (z^3 - w)(z - w)^3 < 0$
$\Rightarrow$ *since* $z \in [0, w]$ *that* $(z^3 - w) > 0$ *and* $z < w$
$\Rightarrow z > \sqrt[3]{w}$ *and* $z < w \Rightarrow s^y - 1 > \sqrt[3]{s - 1}$ *and* $s^y - 1 < s - 1$
$\Rightarrow y > \log_s(1 + \sqrt[3]{s - 1})$ *and* $y < 1$
$\Rightarrow \log_s(1 + \sqrt[3]{s - 1}) < 1 \Rightarrow 1 + \sqrt[3]{s - 1} < s \Rightarrow \sqrt[3]{s - 1} < s - 1$
$\Rightarrow s - 1 > 1 \Rightarrow s > 2$ □

Figures 3.7 and 3.8 show plots of the values of $m_x(\emptyset)$ where y (as defined in theorem 48) varies between $[0, 1]$ and $s = 0.5$ and $s = 40$ respectively.

THEOREM 49 *If $|LA| \leq 3$ then the mass selection function generated by a Frank's t-norm with parameter s (i.e. $f_{\wedge,s}$) is consistent for all $s \in [0, 1]$*
Proof
We first prove the result for $|LA| = 3$
For $x \in \Omega$ let $\mu_{L_1}(x) = y_1$, $\mu_{L_2}(x) = y_2$, $\mu_{L_3}(x) = y_3$. Without loss of generality we need only consider the following four subsets: $\{L_1, L_2, L_3\}$, $\{L_1, L_2\}$, $\{L_1\}$, $\emptyset$. The $s = 1$ and $s = 0$ cases are already proved since these correspond to the consonant and independent msf respectively. Therefore, we assume $s \in (0, 1)$.

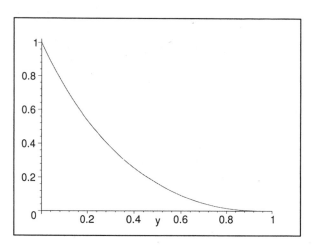

Figure 3.7: Plot of values of $m_x(\emptyset)$ where $s - 0.5 \; \mu_{L_1}(x) - \mu_{L_2}(x) = \mu_{L_3}(x) = y$ and y varies between 0 and 1

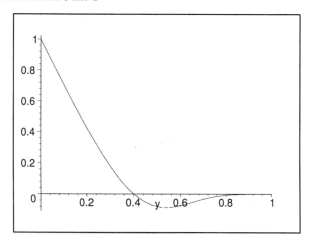

Figure 3.8: Plot of values of $m_x(\emptyset)$ where $s = 40 \; \mu_{L_1}(x) = \mu_{L_2}(x) = \mu_{L_3}(x) = y$ and y varies between 0 and 1

Now by the inversion formula we have:

$$m_x(\{L_1, L_2, L_3\}) = \wedge_s(y_1, y_2, y_3) =$$

$$\log_s\left(1 + \frac{(s^{y_1} - 1)(s^{y_2} - 1)(s^{y_3} - 1)}{(s - 1)^2}\right)$$

$$m_x(\{L_1, L_2\}) = \wedge_t(y_1, y_2) - \wedge_s(y_1, y_2, y_3) =$$

$$\log_s\left(\frac{1 + \frac{(s^{y_1} - 1)(s^{y_2} - 1)}{s - 1}}{1 + \frac{(s^{y_1} - 1)(s^{y_2} - 1)(s^{y_3} - 1)}{(s - 1)^2})}\right)$$

$$m_x(\{L_1\}) = y_1 - \wedge_s(y_1, y_2) - \wedge_s(y_1, y_3) + \wedge_s(y_1, y_2, y_3)$$

$$= \log_s \left(\frac{s^{y_1}\left(1 + \frac{(s^{y_1}-1)(s^{y_2}-1)(s^{y_3}-1)}{(s-1)^2}\right)}{\left(1 + \frac{(s^{y_1}-1)(s^{y_2}-1)}{s-1}\right)\left(1 + \frac{(s^{y_1}-1)(s^{y_3}-1)}{s-1}\right)} \right)$$

$$m_x(\emptyset) = 1 - y_1 - y_2 - y_3 + \wedge_s(y_1, y_2) +$$

$$\wedge_s(y_1, y_3) + \wedge_s(y_2, y_3) - \wedge_s(y_1, y_2, y_3)$$

$$= \log_s \left(\frac{s\left(1 + \frac{(s^{y_1}-1)(s^{y_2}-1)}{s-1}\right)\left(1 + \frac{(s^{y_1}-1)(s^{y_3}-1)}{s-1}\right)\left(1 + \frac{(s^{y_2}-1)(s^{y_3}-1)}{s-1}\right)}{s^{y_1}s^{y_2}s^{y_3}\left(1 + \frac{(s^{y_1}-1)(s^{y_2}-1)(s^{y_3}-1)}{(s-1)^2}\right)} \right)$$

Now let $z_1 = s^{y_1} - 1$, $z_2 = s^{y_2} - 1$ *and* $z_3 = s^{y_3} - 1$ *and note that* $z_1, z_2, z_3 \in$
$[s-1, 0]$:
Now $m_x(\{L_1, L_2, L_3\}) \geq 0$ *trivially since* $\wedge_s$ *is a t-norm.*
Since $s \in (0, 1)$ $m_x(\{L_1, L_2\}) \geq 0 \Leftrightarrow$

$$\frac{(s-1)(s-1+z_1 z_2)}{(s^2 - 2s + 1 + z_1 z_2 z_3)} \leq 1$$

$\Leftrightarrow$ *(since the denominator is* ≥ 0) $(s-1)(s-1+z_1 z_2) - s^2 + 2s - 1 - z_1 z_2 z_3 \leq 0$
$\Leftrightarrow$ $z_1 z_2(s - 1 - z_3) \leq 0$ *This holds trivially since* $z_1 \leq 0$, $z_2 \leq 0$ *and*
$(s - 1 - z_3) \leq 0$.
Since $s \in (0, 1)$ $m_x(\{L_1\}) \geq 0 \Leftrightarrow$

$$\frac{(s^2 - 2s + 1 + z_1 z_2 z_3)(z_1 + 1)}{(s - 1 + z_1 z_3)(s - 1 + z_1 z_2)} \leq 1$$

$\Leftrightarrow$ $(s^2 - 2s + 1 + z_1 z_2 z_3)(z_1 + 1) - (s - 1 + z_1 z_3)(s - 1 + z_1 z_2) \leq 0$
$\Leftrightarrow$ *(since the denominator is* ≥ 0) $z_1(1 - s + z_3)(1 - s + z_2) \leq 0$.
This holds trivially since $z_1 \leq 0$, $(1 - s + z_2) \leq 0$ *and* $(1 - s + z_3) \leq 0$. *Since*
$s \in (0, 1)$ $m_x(\emptyset) \geq 0 \Leftrightarrow$

$$\frac{s(s - 1 + z_1 z_2)(s - 1 + z_1 z_3)(s - 1 + z_2 z_3)}{(s - 1)(z_1 + 1)(z_2 + 1)(z_3 + 1)(s^2 - 2s + 1 + z_1 z_2 z_3)} \leq 1$$

$\Leftrightarrow$ *(since the denominator is* ≤ 0)
$s(s-1+z_1 z_2)(s-1+z_1 z_3)(s-1+z_2 z_3) - (s-1)(z_1+1)(z_2+1)(z_3+1)(s^2 - 2s + 1 + z_1 z_2 z_3) \geq 0 \Leftrightarrow (1 - s + z_3)(1 - s + z_2)(1 - s + z_1)(1 - s + z_1 z_2 z_3) \geq 0$
This holds trivially since $(1 - s + z_1) \geq 0$, $(1 - s + z_2) \geq 0$, $(1 - s + z_3) \geq 0$
and $(1 - s + z_1 z_2 z_3) \geq 0$.

   Since when $|LA| = 3$ *the msf is consistent for* $s \in [0, 1]$ *it must follow that*
when $|LA| < 3$ *the msf must also be consistent, otherwise we could extend any*
counter example where $|LA| < 3$ *to the* $|LA| = 3$ *case by setting the remaining*
appropriateness measures to zero. $\square$

3.9 Alternative Mass Selection Functions

In the previous sections of this chapter the mass selection function that we have considered have been inherently linked to t-norms and their dual t-conorms according to the constraints

$$\forall L_1, \ldots, L_k \in LA \; \mu_{L_1 \wedge \ldots \wedge L_k}(x) = f_\wedge (\mu_{L_1}(x), \ldots, \mu_{L_k}(x))$$
$$\forall L_1, \ldots, L_k \in LA \; \mu_{L_1 \vee \ldots \vee L_k}(x) = f_\vee (\mu_{L_1}(x), \ldots, \mu_{L_k}(x))$$

Such relationships need not hold, however, and in this section we investigate mass selection functions for which the appropriateness of conjunctions (disjunctions) of labels cannot be evaluated according to a t-norm (t-conorm). We begin by giving an example of a rather intuitive msf for which conjunctions of labels are not evaluated according to a t-norm and then go on to show that there exist mass selection functions for which even the monotonicity property C2 of t-norms (chapter 2) can be violated.

DEFINITION 50 *Valid Mass Assignments*
$\forall x \in \Omega$ let $\mathcal{M}_x \subseteq \mathcal{M}$ be the set of possible mass assignments m_x that could be selected by a valid mass selection function given the label appropriateness measures $\mu_{L_1}(x), \ldots, \mu_{L_n}(x)$. Formally,

$$\mathcal{M}_x = \left\{ m_x \in \mathcal{M} : \sum_{T: L_i \in T} m_x(T) = \mu_{L_i}(x) \; for \; i = 1, \ldots, n \right\}$$

$\mathcal{M}_x$ can be represented as a subspace of the $2^n - 1$ dimensional convex hull $V = \{\vec{x} \in [0, 1]^{2^n} : \sum_{i=1}^{2^n} x_i = 1\}$ as follows:

DEFINITION 51 *Geometric Representation of $\mathcal{M}_x$*
Let $id : 2^\Omega \rightarrow \{1, \ldots, 2^n\}$ be an enumeration of 2^Ω. For example,

$$\forall T \subseteq LA \; id(T) = 2^n - \sum_{j=1}^n \epsilon_j(T) 2^{n-j} \; where$$

$\epsilon_j(T) = 1$ if $L_j \in T$ and $= 0$ otherwise

Then a geometric representation of $\mathcal{M}_x$ is given by the following convex volume:

$$V(\mathcal{M}_x) =$$

$$\left\{ \vec{m} \in [0, 1]^{2^n} : \sum_{i=1}^{2^n} m_i = 1 \; and \; \sum_{T: L_j \in T} m_{id(T)} = \mu_{L_j}(x) \; for \; j = 1, \ldots, n \right\}$$

Given this geometric model then a rather natural mass selection function is that which selects the mass assignment, the vector representation of which is the centre of mass of the volume $V(\mathcal{M}_x)$. Notice that since, trivially, $V(\mathcal{M}_x)$ is convex this centre of mass solution will be a member of $\mathcal{M}_x$.

DEFINITION 52 *Centre of Mass msf*
$\forall x \in \Omega$ *the centre of mass selection function selects the mass assignment corresponding to the centre of mass of the volume* $V(\mathcal{M}_x)$. *More formally,*

$$\forall T \subseteq LA \; m_x(T) = \frac{\int_{V(\mathcal{M}_x)} m_{id(T)} dV(\mathcal{M}_x)}{\int_{V(\mathcal{M}_x)} dV(\mathcal{M}_x)}$$

EXAMPLE 53 *Let* $LA = \{L_1, L_2\}$ *so that* $id(\{L_1, L_2\}) = 1$, $id(\{L_1\}) = 2$, $id(\{L_2\}) = 3$ *and* $id(\emptyset) = 4$. *Then* $\forall x \in \Omega \; V(\mathcal{M}_x)$ *is given by:*

$$V(\mathcal{M}_x) = \{\langle m_1, \mu_{L_1}(x) - m_1, \mu_{L_2}(x) - m_1, 1 - \mu_{L_1}(x) - \mu_{L_2}(x) + m_1 \rangle$$
$$: m_1 \in [\max(0, 1 - \mu_{L_1}(x) - \mu_{L_2}(x)), \min(\mu_{L_1}(x), \mu_{L_2}(x))]\}$$

Hence, using the centre of mass msf in this case we obtain:

$$m_x(\{L_1, L_2\}) = \frac{\int_{\max(0, 1 - \mu_{L_1}(x) - \mu_{L_2}(x))}^{\min(\mu_{L_1}(x), \mu_{L_2}(x))} m_1 dm_1}{\int_{\max(0, 1 - \mu_{L_1}(x) - \mu_{L_2}(x))}^{\min(\mu_{L_1}(x), \mu_{L_2}(x))} dm_1}$$

$$= \frac{\frac{1}{2} \times \left(\min(\mu_{L_1}(x), \mu_{L_2}(x))^2 - \max(0, 1 - \mu_{L_1}(x) - \mu_{L_2}(x))^2\right)}{\min(\mu_{L_1}(x), \mu_{L_2}(x)) - \max(0, 1 - \mu_{L_1}(x) - \mu_{L_2}(x))}$$

$$= \frac{\min(\mu_{L_1}(x), \mu_{L_2}(x)) + \max(0, 1 - \mu_{L_1}(x) - \mu_{L_2}(x))}{2}$$

Also, since we have only two labels in LA *it also follows that:*

$$\forall x \in \Omega \; \mu_{L_1 \wedge L_2}(x) =$$
$$\frac{\min(\mu_{L_1}(x), \mu_{L_2}(x)) + \max(0, 1 + \mu_{L_1}(x) + \mu_{L_2}(x))}{2}$$

However, it is well known that the function $\frac{\min(a,b) + \max(0, 1 - a - b)}{2}$ *is not a t-norm since it does not satisfy the associativity property C4 (see chapter 2).*

Indeed the range of possible valid mass selection functions is very wide with some satisfying unusual properties. For instance, we now show that there exists mass selection functions that do not always satisfy the standard monotonicity property for conjunctions of labels. In other words, it is possible to find mass selection functions such that in certain cases where $\mu_{L_2}(x) \geq \mu_{L_3}(x)$ the resulting mass assignment means that, for another label L_1, $\mu_{L_1 \wedge L_2}(x) < \mu_{L_1 \wedge L_3}(x)$.

EXAMPLE 54 *Let $LA = \{L_1, L_2, L_3\}$ and for some $x \in \Omega$ and suppose that $\mu_{L_1}(x) = 0.3$, $\mu_{L_2}(x) = 0.4$ and $\mu_{L_3}(x) = 0.5$. To show that mass selections functions need not preserve the monotonicity property we need only show that there is mass assignment in $\mathcal{M}_x$ for which the property does not hold in this case. One such mass assignment is given as follows:*

$m_x(\{L_1, L_2, L_3\}) = 0.1$, $m_x(\{L_1, L_2\}) = 0.2$, $m_x(\{L_1, L_3\}) = 0$,
$m_x(\{L_2, L_3\}) = 0.1$ $m_x(\{L_1\}) = 0$, $m_x(\{L_2\}) = 0$,
$m_x(\{L_3\}) = 0.3$, $m_x(\emptyset) = 0.3$

We can see that this mass assignment is in $\mathcal{M}_x$ since:

$m_x(\{L_1, L_2, L_3\}) + m_x(\{L_1, L_2\}) + m_x(\{L_1, L_3\}) + m_x(\{L_1\})$
$= 0.1 + 0.2 + 0 + 0 = 0.3 = \mu_{L_1}(x)$ *and*
$m_x(\{L_1, L_2, L_3\}) + m_x(\{L_1, L_2\}) + m_x(\{L_2, L_3\}) + m_x(\{L_2\})$
$= 0.1 + 0.2 + 0.1 + 0 = 0.4 = \mu_{L_2}(x)$ *and*
$m_x(\{L_1, L_2, L_3\}) + m_x(\{L_1, L_3\}) + m_x(\{L_2, L_3\}) + m_x(\{L_3\})$
$= 0.1 + 0 + 0.1 + 0.3 = 0.5 = \mu_{L_3}(x)$

From this mass assignment we have that:

$\mu_{L_1 \wedge L_2}(x) = m_x(\{L_1, L_2, L_3\}) + m_x(\{L_1, L_2\}) = 0.1 + 0.2 = 0.3$ *and*
$\mu_{L_1 \wedge L_3}(x) = m_x(\{L_1, L_2, L_3\}) + m_x(\{L_1, L_3\}) = 0.1 + 0 = 0.1$

Hence, $\mu_{L_1 \wedge L_2}(x) > \mu_{L_1 \wedge L_3}(x)$ even though $\mu_{L_2}(x) < \mu_{L_3}(x)$

3.10 An Axiomatic Approach to Appropriateness Measures

In the previous sections of this chapter we have attempted to use what Giles [35] refers to as a semantic approach in order to define a calculus for appropriateness measures. Such an approach requires us to provide an operational interpretation of the underlying measure and then investigate what emergent calculus arise which are consistent with this interpretation. Alternatively, it is also possible to adopt an axiomatic approach where we begin by identifying a number of intuitive properties that should be satisfied by appropriateness measures and then investigate the nature of the measures characterised by these axioms.

In this section we propose an axiomatic foundation for measures of appropriateness and investigate its relationship to the label semantics model proposed above. The following definition proposes a system of axioms for appropriateness measures on $LE \times \Omega$:

DEFINITION 55 *Axioms for Appropriateness Measures*
An appropriateness measure on $LE \times \Omega$ is a function $\mu : LE \times \Omega \to [0, 1]$

such that $\forall x \in \Omega, \; \forall \theta \in LE \; \mu_\theta(x)$ quantifies the appropriateness of label expression θ as a description of value x and satisfies:

AM1 $\forall \theta \in LE$ *if* $\models \neg\theta$ *then* $\forall x \in \Omega \; \mu_\theta(x) = 0$

AM2 $\forall \theta, \varphi \in LE$ *if* $\theta \equiv \varphi$ *then* $\forall x \in \Omega \; \mu_\theta(x) = \mu_\varphi(x)$

AM3 $\forall \theta \in LE$ *there exists a function* $f_\theta : [0,1]^n \to [0,1]$ *such that*
$\forall x \in \Omega \; \mu_\theta(x) = f_\theta(\mu_{L_1}(x), \ldots, \mu_{L_n}(x))$

AM4 $\forall \theta, \varphi \in LE$ *if* $\models \neg(\theta \wedge \varphi)$ *then* $\forall x \in \Omega \; \mu_{\theta \vee \varphi}(x) = \mu_\theta(x) + \mu_\varphi(x)$

AM1 states that You will always judge a contradiction as an inappropriate description. We might argue for this preservation of the law of contradiction in terms of information content as follows: You view that a contradiction such as 'tall or not tall' can never be an appropriate description of an instance (person) since it contains no information that can help You or another agent identify that instance. For example, suppose You are witness to a robbery and are asked to provide the police with a description of the suspect. Further suppose that the description You provide is as follows:

'He was tall but not tall, with medium but not medium build and his eyes were strangely blue but not blue.'

To what degree does this statement help the police to identify the suspect or at least to eliminate people from their inquiry? We would claim not at all. From a graded-truth perspective it might be argued that 'tall but not tall' identifies an intermediate region of heights around the 'fuzzy' boundary between tall and not tall. However, it is unclear that in practice the police would ever make such an inference and even if they did then it must certainly be doubtful that it would be admissible as evidence in a court of law.

AM2 corresponds to the assumption that classical equivalence is preserved by appropriateness measures. In other words, You view two classically equivalent sentences as having the same meaning and hence as sharing the same level of appropriateness. From the discussion regarding Elkan's paper [30] in chapter 2 it is clear that the question of what equivalences are satisfied by vague concepts is a very difficult one. Hence, since there is no real reason a priori to reject any classical equivalence then it is perhaps reasonable, as a simplification, to assume that all classical equivalences are satisfied.

AM3 is a functionality axiom according to which the appropriateness measure of any logical combination of labels can be evaluated directly from the appropriateness of the labels themselves.

AM4 places some constraints on the form of the functional mappings described in **AM3**. Specifically, in the case of two logically disjoint expressions θ and φ

$$\forall \langle a_1, \ldots, a_n \rangle \in [0,1]^n \; f_{\theta \vee \varphi}(a_1, \ldots a_n) = f_\theta(a_1, \ldots a_n) + f_\varphi(a_1, \ldots a_n)$$

A possible justification for this is as follows: Since You need not take into account any logical dependencies between θ and φ then You can evaluate the appropriateness of $\theta \vee \varphi$ simply by adding the appropriateness measures of θ and φ respectively. A generalisation of this axiom will be investigated in chapter 8.

We now show that this system of axioms can be characterised by the label semantics model introduced in the previous sections of this chapter, taken in conjunction with a mass selection function.

DEFINITION 56 *Logical Atoms*

(i) *Let ATT denote the set of logical atoms of LE (i.e. all expressions of the form $\alpha = \bigwedge_{i=1}^{n} \pm L_i$ where $+L$ denotes L and $-L$ denotes $\neg L$)*

(ii) *Let $ATT_\theta = \{\alpha \in ATT | \alpha \models \theta\}$*

(iii) *$\forall T \subseteq LA \; \alpha_T = \left(\bigwedge_{L_i \in T} L_i\right) \wedge \left(\bigwedge_{L_i \notin T} \neg L_i\right)$*

LEMMA 57 $\forall T \subseteq LA \; \lambda(\alpha_T) = \{T\}$
Proof

$$\forall T \subseteq LA \; \lambda(\alpha_T) = \lambda\left(\left(\bigwedge_{L \in T} L\right) \wedge \left(\bigwedge_{L \notin T} \neg L\right)\right) \quad \text{by definition 56}$$

$$= \lambda\left(\bigwedge_{L \in T} L\right) \cap \lambda\left(\bigwedge_{L \in T^c} \neg L\right) \quad \text{by definition 25}$$

$$= \lambda\left(\bigwedge_{L \in T} L\right) \cap \lambda\left(\neg\left(\bigvee_{L \in T^c} L\right)\right) \quad \text{by corollary 32 and de Morgan's laws}$$

$$= \lambda\left(\bigwedge_{L \in T} L\right) \cap \lambda\left(\bigvee_{L \in T^c} L\right)^c \quad \text{by definition 25}$$

$$= \{R : T \subseteq R\} \cap \{R : T^c \cap R \neq \emptyset\}^c \text{ by definition 25}$$
$$= \{R : T \subseteq R\} \cap \{R : T^c \cap R = \emptyset\} = \{T\} \; \square$$

LEMMA 58

$$\forall \theta \in LE \; ATT_\theta = \{\alpha_T : T \in \lambda(\theta)\}$$

Proof
$(\Rightarrow)$
Let $v_\alpha \in Val$ be defined such that $\forall L \in LA \; v_\alpha(L) = 1$ if and only if $\alpha \models L$
Suppose $\alpha \in ATT_\theta$ then $v_\alpha(\theta) = 1 \Rightarrow \tau(v_\alpha) \in \{\tau(v) : v(\theta) = 1e\}$

$\Rightarrow \tau(v_\alpha) \in \lambda(\theta)$ *by lemma 30*
Now letting $T = \tau(v_\alpha)$ then $\alpha = \alpha_T$ and therefore, $\alpha \in \{\alpha_T : T \in \lambda(\theta)\}$
$(\Leftarrow)$
Suppose $\alpha = \alpha_T$ for some $T \in \lambda(\theta)$ then
$\exists v \in Val : v(\theta) = 1$ *and $T = \tau(v)$ by lemma 30*
$\Rightarrow \alpha \in ATT_\theta$ *since $v(\theta) = 1$ if and only if $\alpha_{\tau(v)} \in ATT_\theta$* $\square$

DEFINITION 59 *For $\Psi \subseteq 2^{LA}$ $\theta_\Psi = \bigvee_{T \in \Psi} \alpha_T$ where α_T is as defined in definition 56*

THEOREM 60

$$\forall \Psi \subseteq 2^{LA} \; \lambda(\theta_\Psi) = \Psi$$

Proof

$$\forall \Psi \subseteq 2^{LA} \lambda(\theta_\Psi) = \lambda \left(\bigvee_{T \in \Psi} \alpha_T \right) \text{ by definition 59}$$

$$= \bigcup_{T \in \Psi} \lambda(\alpha_T) = \bigcup_{T \in \Psi} \{T\} = \Psi \text{ by definition 25 and lemma 57} \;\square$$

Given theorem 60 we see that, in effect, definition 59 provides an inverse to the λ-mapping of definition 25. Up to logical equivalence, this identifies a 1-1 correspondence between label expressions in LE and subsets of 2^{LA}.

EXAMPLE 61

Let $LA = \{small(s), \; medium(m), \; large(l)\}$ then

$\alpha_{\{s, \, m\}} = small \wedge medium \wedge \neg large$

$\alpha_{\{m\}} = medium \wedge \neg small \wedge \neg large$

and therefore if $\Psi = \{\{s, m\}, \{m\}\}$ then

$\theta_\Psi = \alpha_{\{s,m\}} \vee \alpha_{\{m\}} = (small \wedge medium \wedge \neg large) \vee$

$(medium \wedge \neg small \wedge \neg large) \equiv medium \wedge \neg large$

THEOREM 62 *Characterization Theorem*
μ is an appropriateness measure on $LE \times \Omega$ satisfying AM1-AM4 iff $\forall x \in \Omega$ there exists a mass assignment m_x on 2^{LA} such that

$$m_x = \Delta(\mu_{L_1}(x), \ldots, \mu_{L_n}(x))$$

for some mass selection function Δ and

$$\forall \theta \in LE \; \mu_\theta(x) = \sum_{T \in \lambda(\theta)} m_x(T)$$

Proof

(⇐)

If $\models \theta$ *then* $\theta \equiv \varphi \vee \neg\varphi$ *for any* $\varphi \in LE$ *and therefore by corollary 32 and definition 25*

$$\mu_\theta(x) = \sum_{T \subseteq LA} m_x(T) = 1$$

and hence AM1 holds.

$\forall \theta, \varphi \in LE : \theta \equiv \varphi$ *we have by corollary 32 that*

$$\forall x \in \Omega \; \mu_\theta(x) = \sum_{T \in \lambda(\theta)} m_x(T) = \sum_{T \in \lambda(\varphi)} m_x(T) = \mu_\varphi(x)$$

Hence, AM2 holds.

If $\models \neg (\theta \wedge \varphi)$ *then by theorem 33 and definition 25* $\lambda(\theta \wedge \varphi) = \emptyset \Rightarrow \lambda(\theta) \cap \lambda(\varphi) = \emptyset$ *hence* $\forall x \in \Omega$

$$\mu_{\theta \vee \varphi}(x) = \sum_{T \in \lambda(\theta) \cup \lambda(\varphi)} m_x(T) = \sum_{T \in \lambda(\theta)} m_x(T) + \sum_{T \in \lambda(\varphi)} m_x(T) = \mu_\theta(x) + \mu_\varphi(x)$$

Hence AM3 holds.

$\forall \theta \in LE \; \mu_\theta(x) = f_\theta(\mu_{L_1}(x), \ldots, \mu_{L_n}(x))$ *where*

$$f_\theta(\mu_{L_1}(x), \ldots, \mu_{L_n}(x)) = \sum_{T \in \lambda(\theta)} \Delta(\mu_{L_1}(x), \ldots, \mu_{L_n}(x))(T)$$

Hence AM4 holds.

(⇒)

By the disjunctive normal form theorem for propositional logic it follows that $\forall \theta \in LE \; \theta \equiv \bigvee_{\alpha \in ATT_\theta} \alpha$ *therefore by AM2 and AM3 we have that:*

$$\forall x \in \Omega \; \mu_\theta(x) = \mu_{\left(\bigvee_{\alpha \in ATT_\theta} \alpha\right)}(x) = \sum_{\alpha \in ATT_\theta} \mu_\alpha(x)$$

Now $\forall x \in \Omega, \forall T \subseteq LA$ *let*

$$m_x(T) = \mu_{\alpha_T}(x)$$

then clearly m_x *defines a mass assignment on* 2^{LA}. *Also* m_x *can be determined uniquely from* $\mu_{L_1}(x), \ldots, \mu_{L_n}(x)$ *according to the mass selection function*

$$\Delta(\mu_{L_1}(x), \ldots, \mu_{L_n}(x))(T) = f_{\alpha_T}(\mu_{L_1}(x), \ldots, \mu_{L_n}(x))$$

where $f_{\alpha_T} : [0,1]^n \to [0,1]$ *is the function identified for* α_T *by AM4.*
Finally, by lemma 58

$$\forall \theta \in LE \; \mu_\theta(x) = \sum_{\alpha \in ATT_\theta} \mu_\alpha(x) = \sum_{T \in \lambda(\theta)} \mu_{\alpha_T}(x) = \sum_{T \in \lambda(\theta)} m_x(T)$$

as required. □

Theorem 62 demonstrates the fundamentally probabilistic nature of label semantics since for a fixed value of x axioms AM1, AM2 and AM4 ensure that μ corresponds to a probability measure on LE. Indeed we could interpret $\mu_\theta (x)$ as Your subjective probability that θ is appropriate to describe x, or alternatively that θ is appropriate given value (or instance) x. This of course also links label semantics to the likelihood interpretation of fuzzy memberships proposed by Hisdal [44] and described in chapter 2. This relationship will be particularly apparent in chapter 4 when we investigate what kind of probabilistic constraint statements such as 'x is θ' impose on x. Interestingly, from the proof of theorem 62 we see that the mass assignment value $m_x (T) : T \subseteq LA$ can be interpreted as the probability that the atom α_T is appropriate to describe x. The mass assignment representation of appropriateness measures, however, has a number of advantages over the strictly probabilistic representation, principal amongst which is that it allows for a more intuitive treatment of the functionality property as characterized by axiom AM3. For example, it would seem hard to find a intuitive representation of the consonant mass selection function in terms of probabilities on atoms.

3.11   Label Semantics as a Model of Assertions

Up to this point we have viewed appropriateness measures and their underlying mass assignments as a mechanism by which You can assess the level of appropriateness of label expressions for describing some instance $x \in \Omega$. However, given an element x about which You wish to convey information to another intelligent agent You are faced with a much more specific decision problem. What expression $\theta \in LE$ do You choose to assert in order to describe x and how do You use Your appropriateness measure to guide that choice? In this section we shall propose a possible algorithm, based on the mass assignment m_x, which You might adopted in order to restrict the class of possible assertions.

We begin by assuming that Your aim when making an assertion is to be as truthful as possible while also trying to convey as much information as possible. In this case, given a level of belief m_x in what constitutes the set of appropriate labels for x (i.e. what is the value of $\mathcal{D}_x$) then one possible strategy for You to adopt would be to select a threshold δ, where $1 - \delta$ is an acceptable level of doubt, and search for the smallest sets of label sets $\Psi \subseteq 2^{LA}$ for which the aggregated mass across Ψ exceeds δ (i.e. $\sum_{T \in \Psi} m_x (T) \geq \delta$). In other words, You would try to provide the most specific information possible about the value of $\mathcal{D}_x$ that is consistent with Your desire to be truthful (at least to degree δ). Having found such a set of label sets $\Psi \subseteq 2^{LA}$ then in order to convey the information that $\mathcal{D}_x \in \Psi$ as a linguistic assertion, it is necessary for

You to assert some $\varphi \in LE$ for which the corresponding lambda set is Ψ (i.e. $\lambda(\varphi) = \Psi$). From theorem 60 it follows that $\varphi \equiv \theta_\Psi$, the mapping of Ψ into LE as given in definition 59. Applying this θ-mapping to the sets of label sets Ψ that You have identified as providing the most specific reliable information regarding $\mathcal{D}_x$, will result in the set of disjunctive normal forms, DNF_x, of all assertible expressions describing x. In theory all other assertible expression $\mathcal{A}_x$ should then be contained in the set of expressions $\varphi \in LE$ that are equivalent to some expression in DNF_x.

DEFINITION 63 $\mathcal{S}_x$ *is a set of subsets of* 2^{LA} *such that* $\Psi \in \mathcal{S}_x$ *if and only if* $\sum_{T:T\in\Psi} m_x(T) \geq \delta$ *and for any other subset of* 2^{LA}*:* $\Psi' \subset \Psi$ *it holds that* $\sum_{T:T\in\Psi'} m_x(T) < \delta$

DEFINITION 64 *Disjunctive Normal Form Assertions*

$$DNF_x = \{\theta_\Psi : \Psi \in \mathcal{S}_x\}$$

where θ_Ψ *is defined as in definition 59*

DEFINITION 65 *Assertions*

$$\mathcal{A}_x = \{\varphi \in LE : \varphi \equiv \varphi' \text{ and } \varphi' \in DNF_x\}$$

Given that Your general desire is to be truthful and informative then an alternative perspective that You might adopt regarding the assertion decision problem is as follows: It is desirable to assert some $\varphi \in LE$ for which the appropriateness measure $\mu_\varphi(x) \geq \delta$ but which is as specific as possible. This latter condition can be interpreted as meaning that for any $\varphi' \models \varphi$ then $\mu_{\varphi'}(x) < \delta$. The following theorem shows that the two approaches to identifying the set of assertible expressions are in fact equivalent.

THEOREM 66

$$\mathcal{A}_x = \{\varphi \in LE : \mu_\varphi(x) \geq \delta \text{ and } \forall \varphi' \models \varphi \ \mu_{\varphi'}(x) < \delta\}$$

Proof
Suppose $\varphi \in \mathcal{A}_x$ *then* $\varphi \equiv \theta_\Psi$ *for some* $\Psi \in \mathcal{S}_x$ *therefore*

$$\mu_\varphi(x) = \mu_{\theta_\Psi}(x) \text{ (by corollary 32)} = \sum_{T:T\in\lambda(\theta_\Psi)} m_x(T)$$

$$= \sum_{T:T\in\Psi} m_x(T) \text{ (by theorem 60)} \geq \delta \text{ since } \Psi \in \mathcal{S}_x$$

Now suppose that $\exists \varphi' \models \varphi$ *such that* $\mu_{\varphi'}(x) \geq \delta$ *then* $\sum_{T:T\in\lambda(\varphi')}(x) \geq \delta$. *Also by theorem 31 it follows that* $\lambda(\varphi') \subseteq \lambda(\varphi) = \Psi$. *This a contradiction* *since $\Psi \in S_x$ and hence $\forall \varphi' \models \varphi \; \mu_{\varphi'}(x) < \delta$ as required.*

Alternatively suppose for $\varphi \in LE$ it holds that $\mu_\varphi(x) \geq \delta$ and $\forall \varphi' \models \varphi \; \mu_{\varphi'}(x) < \delta$. Trivially, then

$$\sum_{T:T\in\lambda(\varphi)} m_x(T) \geq \delta$$

Now suppose $\exists \; \Psi \subseteq 2^{LA}$ such that $\Psi \subseteq \lambda(\varphi)$ and for which

$$\sum_{T:T\in\Psi} m_x(T) \geq \delta$$

In this case, since $\lambda(\theta_\Psi) = \Psi$ by theorem 60, it holds that $\lambda(\theta_\Psi) \subseteq \lambda(\varphi)$ and therefore by theorem 31 $\theta_\Psi \models \varphi$. Since $\mu_{\theta_\Psi}(x) = \sum_{T:T\in\Psi} m_x(T)$ then this is a contradiction and hence $\lambda(\varphi) \in S_x$. From this it immediately follows that $\theta_{\lambda(\varphi)} \in DNF_x$. Finally, since by corollary 32 $\theta_{\lambda(\varphi)} \equiv \varphi$ it follows that $\varphi \in A_x$ as required. $\square$

EXAMPLE 67 *Suppose $LA = \{red(r), green(g), blue(b)\}$ such that for some element $x \in \Omega$ we have the following appropriateness degrees*

$$\mu_{red}(x) = 0.6, \; \mu_{green}(x) = 0.4, \; \mu_{blue}(x) = 0.2$$

Then assuming the consonant mass selection function we have the following mass assignment m_x:

$$m_x(\{r,g,b\}) = 0.2, \; m_x(\{r,g\}) = 0.2, \; m_x(\{r\}) = 0.2, \; m_x(\emptyset) = 0.4$$

Then letting $\delta = 0.6$ gives

$$S_x = \{\{\emptyset, \{r\}\}, \{\emptyset, \{r,g\}\}, \{\emptyset, \{r,g,b\}\}, \{\{r\}, \{r,g\}, \{r,g,b\}\}\}$$

and hence

$$DNF_x = \{[(\neg r \wedge \neg g \wedge \neg b) \vee (r \wedge \neg g \wedge \neg b)],$$
$$[(\neg r \wedge \neg g \wedge \neg b) \vee (r \wedge g \wedge \neg b)]$$
$$[(\neg r \wedge \neg g \wedge \neg b) \vee (r \wedge g \wedge b)],$$
$$[(r \wedge \neg g \wedge \neg b) \vee (r \wedge g \wedge \neg b) \vee (r \wedge g \wedge b)]\}$$

From this it follows that, for example

$$\neg(g \vee b) \in A_x \text{ since } \neg(g \vee b) \equiv \neg g \wedge \neg b \equiv$$
$$[(\neg r \wedge \neg g \wedge \neg b) \vee (r \wedge \neg g \wedge \neg b)] \text{ and}$$
$$r \wedge (\neg b \vee g) \in A_x \text{ since } r \wedge (\neg b \vee g) \equiv$$
$$[(r \wedge \neg g \wedge \neg b) \vee (r \wedge g \wedge \neg b) \vee (r \wedge g \wedge b)]$$

The fundamental unresolved problem with this approach is that while DNF_x will be finite, $\mathcal{A}_x$ will be infinite. Given that asserting an expression in disjunctive normal form is likely to be rather unnatural, it is clear then that some additional requirement is needed in order to choose between the expressions in $\mathcal{A}_x$. It is rather unclear what form such additional assumptions about assertibility should take and there is obviously a wide range of possibilities. However, for the purpose of illustration we propose one such mechanism for choosing between the members of $\mathcal{A}_x$. It would seem reasonable to suggest that humans find it easier to understand an expression the fewer logical connectives it contains. For example intuitively, the meaning of 'red and green' would seem clearer than the meaning of the logically equivalent expression 'red and green and blue, or red and green and not blue'. Hence, one way of restricting the choice of possible assertion would be to only consider those members of $\mathcal{A}_x$ with minimal numbers of connectives. Interestingly, in example 67 there is only one such expression in $\mathcal{A}_x$ (up to idempotence), this being $\neg (green \vee blue)$. Alternatively, if the threshold δ is dropped to 0.4 then according to this method You would assert $red \wedge green$ (or its idempotent equivalent $green \wedge red$).

3.12 Relating Label Semantics to Existing Theories of Vagueness

The focus of label semantics on the crisp decision problems associated with the use of labels and the underlying uncertainty associated with these decisions, means that it is somewhat related to a number of alternative theories of vague concepts proposed in the literature.

The epistemic theory of vagueness corresponds to the assumption that, for vague concepts, there exists a crisp boundary between instances that satisfy the concept and those that do not, but that the location of this boundary is uncertain. This idea would appear to be consistent with the notion of an uncertain boundary between those values of x for which a label is appropriate and those values for which it is not. Although, it is perhaps more natural to think of the epistemic theory of vagueness as implying the existence of uncertainty regarding the boundaries of the extension of a vague concept, making it more in keeping with the standard random set interpretation of fuzzy sets as proposed by Goodman and Nguyen [38] and discussed in chapter 2.

The interpretation of vagueness in terms of ignorance of precise boundaries, while controversial, has been defended at length by Williamson [107], [108] and Sorensen [98]. The former provides an extensive treatment of the issue arguing against many of the standard objections to the epistemic view. Williamson then advocates a particular version of the theory based on the idea of margins of error. More specifically, it is assumed that we can only 'know' that a certain instance satisfies a concept if all similar instances satisfy the concept. An alternative formulation of 'margin of error' particularly appropriate to the treatment of vagueness is as follows: We can 'know' that an instance satisfies a concept providing that were the concept to be change slightly (perhaps by marginally changing the boundaries) the instance would still satisfy this new updated concept. In other words, a margin of error must be assumed at the level of concept definition. An excellent summary of Williamsons idea and especially the 'margin of error' theory is given in Sainsbury [91].

As well as philosophical arguments there are also some empirical results supporting the epistemic theory. For example, Bonini etal. [12] describe a number of psychological experiments providing some evidence for the epistemic theory. In particular, several tests were carried out where a population was divided into two sets, 'truth-judgers' and 'falsity-judgers'. The former were then asked to select the smallest value x for which a concept (e.g. tall) holds while the latter were asked to pick the largest value y for which the same concept does not hold. In general it was found that $x >> y$ and interestingly, that similar gaps in knowledge exist for vague predicate (e.g. tall) as for a crisp concept relating to an unknown quantity (e.g. above average).

Label semantics is also conceptually linked to the model of assertability as proposed by Alice Kyburg [57] although the underlying calculus is somewhat different. Kyburg proposes a measure of applicability of an expression θ to an instance x, denoted $A_{\theta,x}$, interpreted as the subjective probability that You will judge θ as applicable when describing x. Clearly, the intuitive idea here is strongly related to that of appropriateness measures. Kyburg then goes on to suggest a Bayesian model to determine how Your assertion that 'x is θ' changes the state of knowledge of another intelligent agent that is not dissimilar from that subsequently proposed for label semantics in chapter 4. However, Kyburg then goes on to make some rather strange assumptions implicity effecting the functionality of the applicability measure $A_{\theta,x}$. For instance, if You make the dual assertion that 'x is θ and x is φ', in effect that 'x is $\theta \wedge \varphi$', then it is assumed that 'each judgment in such pairs is made independently of the other'. It is even suggested that such an independence assumption might hold in the case of the assertion 'x is θ and x is not θ'. Clearly, such an assumption will result in a truth-functional calculus where

$$\forall \theta, \varphi \ A_{\theta \wedge \varphi, x} = A_{\theta,x} \times A_{\varphi,x}$$

However, it is difficult to imagine why any intelligent agent should make judgements regarding the applicability of θ and φ independently, without taking into account the meaning of these concepts.

Another view of vague concepts related to label semantics is given by Rohit Parikh in [77] where he argues for a so-called 'anti-representational' view of vagueness, focusing on the notion of assertibility rather than that of truth. Parikh argues that it is almost unavoidable that different speakers will use the same predicate in different ways because of the manner in which language is learnt. Since vague expressions lack a clear definition we tend to learn the 'usage of these words in some few cases and then we extrapolate. These extrapolations will agree to some extent since we humans are rather similar to each other. But we are not exactly the same.' Certainly it would also be likely that if our subjective knowledge about the assertibility or appropriateness of words is acquired as a result of such a process of extrapolation then we would expect such knowledge to be fundamentally uncertain. This observation would certainly seem consistent with the label semantics approach. There does seem to be some difference, however, between Parikh's notion of assertibility and the notion of appropriateness in label semantics, in that whereas the latter is essentially subjective the former seems to be inter-subjective. By this we mean that the appropriateness of expressions in label semantics is a decision made by individual agents whereas Parikh's assertibility can only be judged across a population. Hence, the variability of use of words across a population would mean that there is no clear dividing line between their assertibility and non-assertibility. To quote Parikh directly:

'Certain sentences are assertible in the sense that we might ourselves assert them and other cases of sentences which are non-assertible in the sense that we ourselves (and many others) would reproach someone who used them. But there will also be the intermediate kind of sentences, where we might allow their use'

Of course it may be possible to link these two notions by supposing that an individual agent's decisions on the appropriateness of labels might well be based on his/her knowledge of this collective property of assertibility.

Summary

In this chapter we have introduced label semantics as an alternative approach to modelling vague concepts. Instead of attempting to explicitly represent the meaning of such concepts in terms of their extensions the focus is on the decision problem faced by an intelligent agent (You) when identifying what labels can be appropriately used to describe an element $x \in \Omega$, as part of an assertion aimed at informing another agent. For $x \in \Omega$ a mass assignment m_x is defined as representing Your belief regarding what constitutes the set of labels that are appropriate for describing x (denoted $\mathcal{D}_x$). From this mass assignment a measure of appropriateness for any expression θ as a description of x, denoted $\mu_\theta(x)$, can then be determined by summing m_x across $\lambda(\theta) \subseteq 2^{LA}$, defined as those values of $\mathcal{D}_x$ consistent with the assertion 'x is θ'.

In the context of this basic framework we have investigated a number of possible calculi for combining appropriateness measures. All such calculi satisfy the standard Boolean properties while also being functional, though never truth functional. The assumption of functionality for appropriateness measures in fact corresponds to the assumption of a particular mass selection function as a mechanism for selecting a single mass assignment from all those consistent with the appropriateness measure values on the basic labels in LA. Interestingly, the nature of the mass selection function can be linked directly to the behaviour of the associated appropriateness measure across conjunctions of labels. Given this relationship we also investigated the consistency of particular t-norms with the label semantics framework. However, not all mass selection functions are linked to t-norms and we have shown that msf can be found that violate t-norm axioms such as associativity (C4) and monotonicity (C2). Finally, we have proposed how appropriateness measures might be used to guide the agent in their choice of what assertion to make and we have discussed the relationship between label semantics and other theories of vague concepts.

Notes

1 The phrase 'degree of belief' is used here in a generic way meaning a subjective quantification of uncertainty rather than with reference to any particular formalism such as Dempster-Shafer theory.

Chapter 4

MULTI-DIMENSIONAL AND MULTI-INSTANCE LABEL SEMANTICS

This chapter introduces two natural extensions to the calculus of appropriateness measures proposed in chapter 3. Firstly, multi-dimensional measures are introduced to quantify the appropriateness of expressions for describing different attributes of a given object. Secondly, we consider the linguistic description of a set of objects DB, where the aim is to quantify how appropriate an expression θ is, on average, for describing the elements of DB. Both of these extensions are crucial to the development of label semantics as a practical framework for modelling and reasoning using vague concepts.

4.1   Descriptions Based on Many Attributes

In chapter 3 the calculus for appropriateness measure was developed on the assumption that there was essentially one attribute of the object, taking values in Ω, that needed to be considered when describing it. However, in practice most objects can only really be described in terms of a variety of features. For example, You would tend to describe a tree in terms of its height, its colour, the type of leaves etc. Therefore, if label semantics is to provide an effective knowledge representation framework for linguistic modelling it must be generalised to the multi-dimensional case. In other words, we need to provide a means of interpreting and reasoning with linguistic expressions involving more than one variable.

Specifically, consider the problem of describing an object based on k attributes $x_1, \ldots, x_k$ with associated universes $\Omega_1, \ldots, \Omega_k$. For each attribute we defined a set of labels $LA_j = \{L_{1,j}, \ldots, L_{n_j,j}\}$ for $j = 1, \ldots, k$. In this case the decision problem of how to describe the object will involve You considering what is the set of appropriate labels for describing each attribute $x_j : j = 1, \ldots, k$. Hence, You need to evaluate Your level of belief in what is the value of the vector $\langle \mathcal{D}_{x_1}, \ldots, \mathcal{D}_{x_k} \rangle$ (i.e. what are the set of appropriate labels for each attribute) for the attribute vector $\langle x_1, \ldots, x_k \rangle$. In this context we can extend the definitions of mass assignment and appropriateness measure given in chapter 3 to the multi-dimensional case. For $T_j \subseteq LA_j : j = 1, \ldots, k$ let $m_{\vec{x}}(T_1, \ldots, T_k)$ denote Your level of belief that the set of appropriate labels for attribute x_j is $T_j \subseteq LA_j$ (i.e. that $\mathcal{D}_{x_j} = T_j$) for $j = 1, \ldots, k$. The measure $m_{\vec{x}}$ can be viewed as a mass relation (joint mass assignment) on labels defined as follows:

DEFINITION 68 *Mass Relation on Labels*
A mass relation on labels is a function $m : 2^{LA_1} \times \ldots \times 2^{LA_k} \to [0,1]$ *such that*

$$\sum_{T_1 \subseteq LA_1} \cdots \sum_{T_k \subseteq LA_k} m(T_1, \ldots, T_k) = 1$$

4.2   Multi-dimensional Label Expressions and λ-Sets

Now given the mass relation $m_{\vec{x}}$ You will want to evaluate the appropriateness of certain label expressions θ for describing the vector of attribute values $\vec{x}$. For example, we might want to know how appropriate it is to assert that '(x_1 is *small* and x_2 is *high*) or (x_1 is *large* and x_2 is *low*)' where *small*, *large* $\in LA_1$ and *low*, *high* $\in LA_2$. This can be rewritten as the assertion '$\langle x_1, x_2 \rangle$ is $(small \wedge high) \vee (large \wedge low)$' where the expression $(small \wedge high) \vee (large \wedge low)$ constrains the vector of appropriate label sets $\langle \mathcal{D}_{x_1}, \mathcal{D}_{x_2} \rangle$ to belonging to a subset of $2^{LA_1} \times 2^{LA_2}$ identified by application of a multi-dimensional version of the λ-mapping. The latter will be defined in the following section but we first formally define the set of possible label expressions for describing an attribute vector $\vec{x}$.

Let LE_j be the set of label expression for variable x_j generated by recursive application of the connectives $\wedge, \vee, \to, \neg$ to the labels LA_j according to definition 1. We can now define the set of multi-dimensional label expressions for describing linguistic relationships between variables as follows:

DEFINITION 69 *(Multi-dimensional Label Expressions)*
$MLE^{(k)}$ *is the set of all multi-dimensional label expressions that can be generated from the label expressions* $LE_j : j = 1, \ldots, k$ *and is defined recursively by:*

(i) If $\theta \in LE_j$ *for* $j = 1, \ldots, k$ *then* $\theta \in MLE^{(k)}$

(ii) If $\theta, \varphi \in MLE^{(k)}$ *then* $\neg \theta, \theta \wedge \varphi, \theta \vee \varphi, \theta \to \varphi \in MLE^{(k)}$

Any k−dimensional label expression θ identifies a subset of $2^{LA_1} \times \ldots \times 2^{LA_k}$, denoted $\lambda^{(k)}(\theta)$, constraining the cross product of label descriptions $\mathcal{D}_{x_1} \times \ldots \times \mathcal{D}_{x_k}$. In this way the imprecise constraint θ on $x_1 \times \ldots \times x_k$ is interpreted as the precise constraint $\mathcal{D}_{x_1} \times \ldots \times \mathcal{D}_{x_k} \in \lambda^{(k)}(\theta)$.

DEFINITION 70 *(Multi-dimensional $\lambda-Sets$)*

$\lambda^{(k)} : MLE^{(k)} \rightarrow 2^{\left(2^{LA_1} \times \ldots \times 2^{LA_k}\right)}$ *is defined recursively as follows:*
$\forall \theta \in MLE^{(k)}$ $\lambda^{(k)}(\theta) \subseteq 2^{LA_1} \times \ldots \times 2^{LA_k}$ *such that*

(i) $\forall \theta \in LE_j : j = 1, \ldots, k$ $\lambda^{(k)}(\theta) = \lambda(\theta) \times \times_{i \neq j} 2^{LA_i 1}$

(ii) $\forall \theta, \varphi \in MLE^{(k)}$ $\lambda^{(k)}(\theta \wedge \varphi) = \lambda^{(k)}(\theta) \cap \lambda^{(k)}(\varphi)$

(iii) $\lambda^{(k)}(\theta \vee \varphi) = \lambda^{(k)}(\theta) \cup \lambda^{(k)}(\varphi)$

(iv) $\lambda^{(k)}(\theta \rightarrow \varphi) = \lambda^{(k)}(\theta)^c \cup \lambda^{(k)}(\varphi)$

(v) $\lambda^{(k)}(\neg\theta) = \lambda^{(k)}(\theta)^c$

EXAMPLE 71 *Consider a modelling problem with two variables x_1 and x_2 for which $LA_1 = \{small(s), medium(m), large(l)\}$ and $LA_2 = \{low(lw), moderate(md), high(h)\}$. $\lambda^{(2)}([s \wedge h] \vee [l \wedge lw])$ can be determined recursively according to the rules given in definition 70 as illustrated by the parsing tree shown in figure 4.1. Here the expression is broken down until, in the leaves of the tree, we only have label expressions describing one particular attribute. Then starting from these leaves the λ-set is calculated by applying the recursion rules of definition 70 up through the tree. The pairs of label sets contained in $\lambda^{(2)}([s \wedge h] \vee [l \wedge lw])$ correspond to the grey cells in figure 4.2. This tabular representation can be simplified if we only show the elements of $\mathcal{F}_1$ and $\mathcal{F}_2$, where $\mathcal{F}_i$ is the set of focal elements for $LA_i : i = 1, 2$. For instance, suppose the focal elements for LA_1 and LA_2 are, respectively:*

$$\mathcal{F}_1 = \{\{small\}, \{small, medium\}, \{medium\}, \{medium, large\}, \{large\}\}$$
$$\mathcal{F}_2 = \{\{low\}, \{low, moderate\}, \{moderate\}, \{moderate, high\}, \{high\}\}$$

In this case $\lambda^{(2)}([s \wedge h] \vee [l \wedge lw])$ can be represented by the tableau shown in figure 4.3

4.3 Properties of Multi-dimensional Appropriateness Measures

We now make a conditional independence assumption regarding the way in which Your decision about appropriate labels for a particular attribute is dependent on the values of other attributes. Specifically, it is assumed that Your belief regarding what are the appropriate labels for variable x_j is dependent only on the value of x_j, once this is known, and is independent of the value of any other variables. For example, suppose that an individual is described by their height and their hair colour, then once You know their actual height You have all the information required to assess whether or not they are tall. In other words, if You know their height You do not need to know anything about

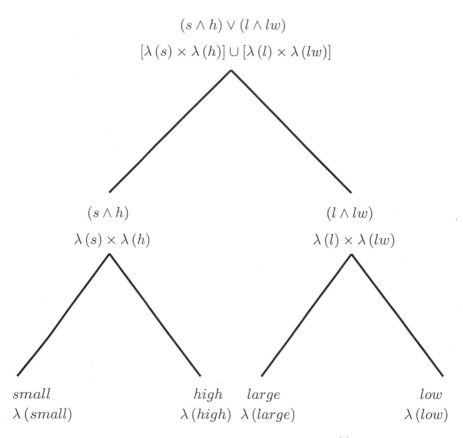

Figure 4.1: Recursive evaluation of the multi-dimensional λ-set, $\lambda^{(2)}$ ($[s \wedge h] \vee [l \wedge lw]$)

their hair colour when making this judgement. This is actually quite a weak assumption and does not *a prior* imply independence between the variables. Even if there are, for instance, strong relationships between those individuals who are tall and those who are blonde, these dependencies do not need to be taken into account when deciding how to describe a particular individual once You know their actual height and hair colour. Given this assumption then we

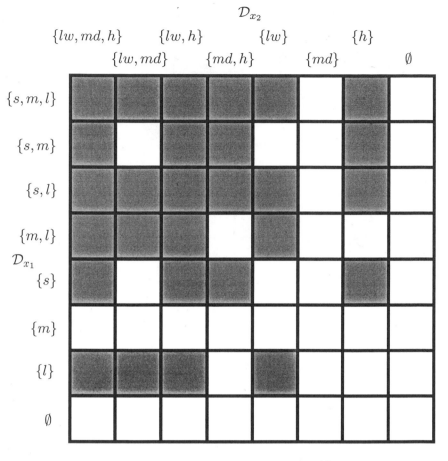

Figure 4.2: Representation of the multi-dimensional λ-set, $\lambda^{(2)}$ ($[s \wedge h] \vee [l \wedge lw]$) as a subset of $2^{LA_1} \times 2^{LA_2}$. The grey cells are those contained within the λ-set.

have that:

$$\forall x_j \in \Omega_j \ \forall T_j \subseteq LA_j : j = 1, \ldots, k \ m_{\langle x_1, \ldots, x_k \rangle}(T_1, \ldots, T_k) =$$

$$\prod_{j=1}^{k} m_{x_j}(T_j)$$

and hence,

$$\forall \theta \in MLE^{(k)} \mu_{\theta}^{(k)}(x_1, \ldots, x_k) = \sum_{\langle T_1, \ldots, T_k \rangle \in \lambda^{(k)}(\theta)} \prod_{j=1}^{k} m_{x_j}(T_j)$$

Given a k dimensional mass relation we can define a projection on to a lower c dimensional mass relation as follows: The project of the mass relation m on

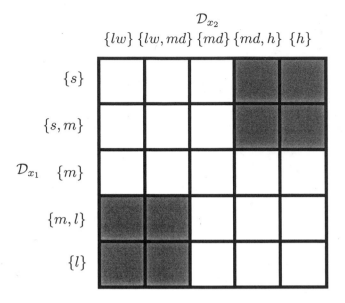

Figure 4.3: Representation of the multi-dimensional λ-set, $\lambda^{(2)}$ ($[s \wedge h] \vee [l \wedge lw]$), showing only the focal cells $\mathcal{F}_1 \times \mathcal{F}_2$. The grey cells are those contained within the λ-set.

$2^{LA_1} \times \ldots 2^{LA_k}$ onto the space $2^{LA_1} \times \ldots \times 2^{LA_c}$ where $c < k$ is given by:

$$m \downarrow_c (T_1, \ldots, T_c) = \sum_{T_{c+1} \subseteq LA_{c+1}} \cdots \sum_{T_k \subseteq LA_k} m (T_1, \ldots, T_k)$$

From this equation we have that the projection of the mass assignment $m_{\langle x_1, \ldots, x_k \rangle}$ onto $2^{LA_1} \times \ldots \times 2^{LA_c}$ is given by:

$$m_{\langle x_1, \ldots, x_k \rangle} \downarrow_c (T_1, \ldots, T_c) =$$

$$\sum_{T_{c+1} \subseteq LA_{c+1}} \cdots \sum_{T_k \subseteq LA_k} \prod_{j=1}^{k} m_{x_j}(T_j) = \prod_{j=1}^{c} m_{x_j}(T_j)$$

Hence the projection of $m_{\langle x_1, \ldots, x_k \rangle}$ onto $2^{LA_1} \times \ldots \times 2^{LA_c}$ for some object with attribute values $x_1, \ldots, x_k$ corresponds to the mass relation on the set of appropriate labels for that object when only the features $x_1, \ldots, x_c$ are taken into consideration. In other words,

$$m_{\langle x_1, \ldots, x_k \rangle} \downarrow_c = m_{\langle x_1, \ldots, x_c \rangle}$$

Now suppose You are interested in quantifying the appropriateness of an expression θ for describing some object where θ relates to a subset of the attributes $x_1, \ldots, x_c$ and makes no reference to the remaining attributes $x_{c+1}, \ldots, x_k$. In other words, $\theta \in MLE^{(c)}$ where $c < k$. In such a case we would expect the value of the k dimensional appropriateness measure $\mu_\theta^{(k)}(x_1, \ldots, x_k)$ to correspond to the appropriateness measure for θ when only the attributes $x_1, \ldots, x_c$ are taken into account (i.e. $\mu_\theta^{(c)}(x_1, \ldots, x_c)$). The following theorem shows that this is indeed the case.

THEOREM 72

If $\theta \in MLE^{(c)}$ for $c < k$ then $\forall x_j \in \Omega_j : j = 1, \ldots, k$

$$\mu_\theta^{(k)}(x_1, \ldots, x_k) = \mu_\theta^{(c)}(x_1, \ldots, x_c)$$

Proof
By definition 70 $\lambda^{(k)}(\theta) = \lambda^{(c)}(\theta) \times \times_{j=c+1}^k 2^{LA_j}$ and therefore,

$$\mu_\theta^{(k)}(x_1, \ldots, x_k) = \sum_{\langle T_1, \ldots, T_c\rangle \in \lambda^{(c)}(\theta)} \sum_{\langle T_{c+1}, \ldots, T_k\rangle \in 2^{LA_{c+1}} \times \ldots \times 2^{LA_k}} \prod_{j=1}^k m_{x_j}(T_j)$$

$$= \sum_{\langle T_1, \ldots, T_c\rangle \in \lambda^{(c)}(\theta)} \prod_{j=1}^c m_{x_j}(T_j) \sum_{T_{c+1} \subseteq LA_{c+1}} \cdots \sum_{T_k \subseteq LA_k} \prod_{j=c+1}^k m_{x_j}(T_j)$$

$$= \sum_{\langle T_1, \ldots, T_c\rangle \in \lambda^{(c)}(\theta)} \prod_{j=1}^c m_{x_j}(T_j) \left[\prod_{j=c+1}^k \sum_{T_j \subseteq LA_j} m_{x_j}(T_j) \right]$$

$$= \sum_{\langle T_1, \ldots, T_c\rangle \in \lambda^{(c)}(\theta)} \prod_{j=1}^c m_{x_j}(T_j) = \mu_\theta^{(c)}(x_1, \ldots, x_c)$$

as required. □

Typically, rule based systems for prediction consist of a knowledge base of conditional rules of the form:
IF $(x_1$ is $\theta_1)$ AND $(x_2$ is $\theta_2)$ AND $\ldots$ AND $(x_{k-1}$ is $\theta_{k-1})$ THEN $(x_k$ is $\theta_k)$
where $x_1, \ldots, x_{k-1}$ are the inputs to the system and x_k is the output. Such rules correspond to a particular family of expressions from $MLE^{(k)}$ of the form:

$$\left(\bigwedge_{j=1}^{k-1} \theta_j \right) \rightarrow \theta_k \text{ where } \theta_j \in LE_j : j = 1, \ldots, k$$

The following theorem shows that the appropriateness measure for expressions of this form can be determined from the appropriateness measures for the individual constraints on each attribute, $\mu_{\theta_j}(x_j) : j = 1, \ldots, k$, by application of the Reichenbach implication operator.

THEOREM 73 *Let* $\theta_j \in LE_j : j = 1, \ldots, k$ *then the appropriateness measure of the conditional* $\left(\bigwedge_{j=1}^{k-1} \theta_j \right) \to \theta_k$ *is given by:*

$$\mu^{(k)}_{(\bigwedge_{j=1}^{k-1} \theta_j) \to \theta_k}(x_1, \ldots, x_k) = 1 - \prod_{j=1}^{k-1} \mu_{\theta_j}(x_j) + \prod_{j=1}^{k} \mu_{\theta_j}(x_j)$$

Proof

By definition 70 we have that:

$$\lambda^{(k)} \left(\left(\bigwedge_{j=1}^{k-1} \theta_j \right) \to \theta_k \right) = \left[\lambda^{(k)}(\bigwedge_{j=1}^{k-1} \theta_j) \right]^c \cup \lambda^{(k)}(\theta_k) =$$

$$\left[\lambda^{(k)}(\bigwedge_{j=1}^{k-1} \theta_j) \cap \left[\lambda^{(k)}(\theta_k) \right]^c \right]^c$$

Now again by definition 70 it follows that:

$$\lambda^{(k)}(\bigwedge_{j=1}^{k-1} \theta_j) \cap \left[\lambda^{(k)}(\theta_k) \right]^c = \lambda(\theta_1) \times \ldots \times \lambda(\theta_{k-1}) \times [\lambda(\theta_k)]^c$$

Therefore,

$$\mu^{(k)}_{(\bigwedge_{j=1}^{k-1} \theta_j) \to \theta_k}(x_1, \ldots, x_k) =$$

$$1 - \left[\prod_{j=1}^{k-1} \sum_{T_j \in \lambda(\theta_j)} m_{x_j}(T_j) \right] \times \left[\sum_{T_k \in [\lambda(\theta_k)]^c} m_{x_k}(T_k) \right]$$

$$= 1 - \left[\prod_{j=1}^{k-1} \mu_{\theta_j}(x_j) \right] \times [1 - \mu_{\theta_k}(x_k)] = 1 - \prod_{j=1}^{k-1} \mu_{\theta_j}(x_j) + \prod_{j=1}^{k} \mu_{\theta_j}(x_j)$$

as required. □

EXAMPLE 74 *Consider a modelling problem with two variables* x_1, x_2 *each with universe* $[0, 10]$ *and for which we have defined the label sets* $LA_1 = \{small_1(s_1), medium_1(m_1),$
$large_1(l_1)\}$ *and* $LA_2 = \{small_2(s_2), medium_2(m_2), large_2(l_2)\}$. *For both variables the appropriateness degrees for* $small, medium$ *and* $large$ *are defined as in example 39. Now suppose we learn that:*

IF x_1 *is medium but not large* THEN x_2 *is medium*

then according to theorem 73 the appropriateness degree for $medium_1 \wedge \neg large_1 \rightarrow medium_2$ is given by:

$$\mu^{(2)}_{medium_1 \wedge \neg large_1 \rightarrow medium_2}(x_1, x_2) =$$

$$1 - \mu_{medium_1 \wedge \neg large_1}(x_1) + \mu_{medium_1 \wedge \neg large_1}(x_1)\mu_{medium_2}(x_2)$$

Assuming the appropriateness degrees for small, medium and large given in example 39 then the resulting function is shown in figure 4.4.

Clearly, this function provides information regarding the relationship between x_1 and x_2 assuming the constraint $medium_1 \wedge \neg large_1 \rightarrow medium_2$. For instance, from figure 4.4 we can see that if $x_1 = 5$ the it is very unlikely that $8 \leq x_2 \leq 10$. The problem of how to make output predictions (for x_2) given input values (for x_1) is considered in detail in a later chapter. Now suppose we

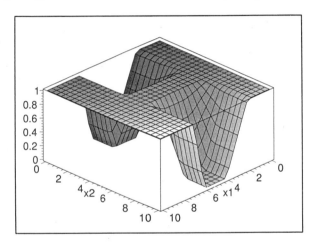

Figure 4.4: Plot of the appropriateness degree for $medium_1 \wedge \neg large_1 \rightarrow medium_2$

also learn that

$$\text{IF } x_1 \text{ is large THEN } x_2 \text{ is small}$$

In this case we would want to evaluate the appropriateness degrees for the expression
$(medium_1 \wedge \neg large_1 \rightarrow medium_2) \wedge (large_1 \rightarrow small_2)$. *For this expression we have:*

$$\lambda^{(2)}((m_1 \wedge \neg l_1 \rightarrow m_2) \wedge (l_1 \rightarrow s_2))$$

$$= \left(\left[\lambda^{(2)}(m_1 \wedge \neg l_1) \right]^c \cup \lambda^{(2)}(m_2) \right) \cap \left(\left[\lambda^{(2)}(l_1) \right]^c \cup \lambda^{(2)}(s_2) \right) =$$

$$\left[\left(\lambda^{(2)}(m_1 \wedge \neg l_1) \cap \left[\lambda^{(2)}(m_2) \right]^c \right) \cup \left(\lambda^{(2)}(l_1) \cap \left[\lambda^{(2)}(s_2) \right]^c \right) \right]^c$$

$$= [(\lambda(m_1 \wedge \neg l_1) \times [\lambda(m_2)]^c) \cup (\lambda(l_1) \times [\lambda(s_2)]^c)]^c$$

Now,

$$(\lambda(m_1 \wedge \neg l_1) \times [\lambda(m_2)]^c) \cap (\mathcal{F}_1 \times \mathcal{F}_2) =$$
$$\{\{s_1, m_1\}, \{m_1\}\} \times \{\{m_2\}, \{m_2, l_2\}, \{l_2\}, \emptyset\}$$

and

$$(\lambda(l_1) \times [\lambda(s_2)]^c) \cap (\mathcal{F}_1 \times \mathcal{F}_2) =$$
$$\{\{l_1\}, \{m_1, l_1\}\} \times \{\{m_2\}, \{m_2, l_2\}, \{l_2\}, \emptyset\}$$

Hence, since these two sets on $2^{LA_1} \times 2^{LA_2}$ are mutually exclusive it follows that:

$$\forall x_1 \in \Omega_1, \forall x_2 \in \Omega_2 \ \mu^{(2)}_{(m_1 \wedge \neg l_1 \to m_2) \wedge (l_1 \to s_2)}(x_1, x_2) = 1-$$

$$\left(\sum_{T_1 \in \lambda(m_1 \wedge \neg l_1)} \sum_{T_2 \in [\lambda(m_2)]^c} m_{x_1}(T_1) m_{x_2}(T_2) + \sum_{T_1 \in \lambda(l_1)} \sum_{T_2 \in [\lambda(s_2)]^c} m_{x_1}(T_1) m_{x_2}(T_2) \right)$$

$$= 1 - (\mu_{m_1 \wedge \neg l_1}(x_1) \times (1 - \mu_{m_2}(x_2)) + \mu_{l_1}(x_1) \times (1 - \mu_{s_2}(x_2)))$$

Again assuming the appropriateness degrees for small, medium and large given in example 39 then the resulting function is shown in figure 4.5.

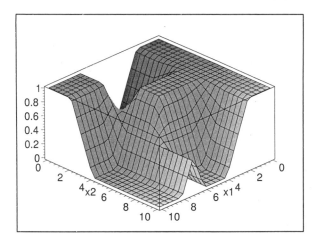

Figure 4.5: Plot of the appropriateness degree for $(medium_1 \wedge \neg large_1 \to medium_2) \wedge (large_1 \to small_2)$

4.4   Describing Multiple Objects

We now investigate another extension of label semantics, in this case to situations where You need to describe more than one object. For simplicity we begin by considering objects described in terms of only one attribute x, with values in universe Ω, and then later extend this to the multi-attribute case. Suppose that You have a set (or database) of objects DB that You wish to describe in terms of LA, the set of labels for values in Ω. One way for You to assess which labels are appropriate to describe the objects in DB would be for You to evaluate a mass assignment m_{DB} where for $T \subseteq LA$, $m_{DB}(T)$ quantifies Your level of belief that object i with attribute value $x(i)$ drawn at random from DB will have the set of appropriate labels T (i.e. that $\mathcal{D}_{x(i)} = T$). Given the interpretation of m_x as a subjective probability distribution for the random set $\mathcal{D}_x$ then m_{DB} is naturally given by:

$$\forall T \subseteq LA \; m_{DB}(T) = \frac{\sum_{i \in DB} m_{x(i)}(T)}{|DB|}$$

Having evaluated m_{DB} we can then calculate the level of appropriateness of any expression θ for describing the elements of DB by first determining $\lambda(\theta)$ according to definition 25, and then summing m_{DB} over this set of label sets. More formally:

$$\forall \theta \in LE \; \mu_\theta(DB) = \sum_{T \in \lambda(\theta)} m_{DB}(T)$$

Effectively, $\mu_\theta(DB)$ quantifies the degree to which on average the elements of DB can be described as being θ.

EXAMPLE 75 *Consider the following set of people*

$$DB = \{Bob, \; Fred, \; Mary, \; Jane, \; Ethel, \; Henry\}$$

each described according to the attribute age with universe $\Omega =$ $[0, 80]$. *Let the set of labels for describing age be* $LA =$ $\{young(y), \; middle \; aged(ma), \; old(o)\}$ *with appropriateness measures given in figure 4.6. Given these appropriateness measures and assuming the consonant msf then the underlying mass assignments can be calculated for specific ages. These are shown in figure 4.7 as age varies across* Ω. *Suppose that for the elements of* DB *the values of age are as follows:*

$$age(Bob) = 25, \; age(Fred) = 55, \; age(Mary) = 31, \; age(Jane) = 28,$$
$$age(Ethel) = 80, \; age(Henry) = 40$$

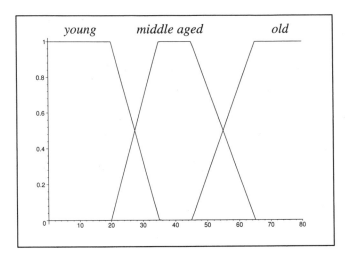

Figure 4.6: Appropriateness measures for labels young, middle aged and old

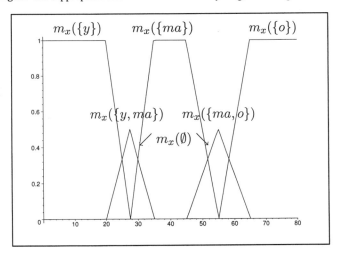

Figure 4.7: Mass assignment values for m_x generated according to the consonant msf as x varies from 0 to 80

Hence, we can calculate m_{DB} as follows:

$$m_{DB}\left(\{y\}\right) = \frac{1}{6}(m_{age(Bob)}\left(\{y\}\right) + m_{age(Fred)}\left(\{y\}\right) +$$

$$m_{age(Mary)}\left(\{y\}\right) + m_{age(Jane)}\left(\{y\}\right)$$

$$+m_{age(Ethel)}\left(\{y\}\right) + m_{age(Henry)}\left(\{y\}\right)) = \frac{1}{18}$$

similarly,

$$m_{DB}\left(\{y, ma\}\right) = \frac{8}{45}, \; m_{DB}\left(\{ma\}\right) = \frac{23}{90}, \; m_{DB}\left(\{y, o\}\right) = \frac{1}{12},$$

$$m_{DB}\left(\{o\}\right) = \frac{1}{6} \; and \; m_{DB}\left(\emptyset\right) = \frac{47}{180}$$

Given m_{DB} You may now wonder to what degree, on average, can the people

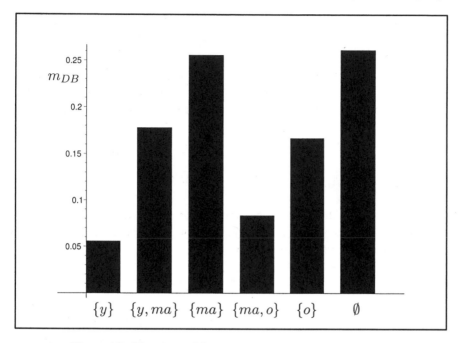

Figure 4.8: Histogram of the aggregated mass assignment m_{DB}

in DB be described as being 'either not old or middle aged'. In other words, what is the value of $\mu_{\neg o \vee ma}(DB)$? To evaluate this we must first determine $\lambda(\neg o \vee ma)$ although in fact it is sufficient to $\lambda(\neg o \vee ma) \cap \mathcal{F}$, where $\mathcal{F}$ is the set of focal elements for LA as given in definition 40. In this case,

$$\mathcal{F} = \{\{y\}, \{y, ma\}, \{ma\}, \{ma, o\}, \{o\}, \emptyset\}$$

Now $\neg o \vee ma \equiv \neg(o \wedge \neg ma)$ and hence by definition 25 and corollary 32 we have that:

$$\lambda(\neg o \vee ma) \cap \mathcal{F} = \lambda(o \wedge \neg ma)^c \cap \mathcal{F} = \{\{y\}, \{y, ma\}, \{ma\}, \{ma, o\}, \emptyset\}$$
$$= \mathcal{F} - \{\{o\}\}$$

Hence,

$$\mu_{\neg o \vee ma}(DB) = 1 - m_{DB}(\{o\}) = 1 - \frac{1}{6} = \frac{5}{6}$$

Multi-object label semantics can also be extended to the multi-attribute case using the ideas introduced in the first sections of this chapter. Suppose that our database corresponds to a set of objects described by attributes $x_1, \ldots, x_k$. In this case we can view DB as a sets of vectors of attribute values, one for each object under consideration. More formally,

$$DB = \{\langle x_1\,(i)\,,\ldots,x_k\,(i)\rangle : i = 1,\ldots,N\}$$

The decision problem associated with describing DB linguistically now requires You to quantify the likelihood that an element selected at random from DB will have the set of labels $T_i \subseteq LA_i$ that are appropriate to describe attribute x_i for $i = 1,\ldots,k$ (i.e. that $\langle \mathcal{D}_{x_1}, \ldots, \mathcal{D}_{x_k}\rangle = \langle T_1, \ldots, T_k\rangle$). This uncertainty measure is given by the following mass relation:

$$\forall T_j : j = 1,\ldots,k \; m_{DB}\,(T_1,\ldots,T_k) = \frac{\sum_{i \in DB} m_{\vec{x}(i)}\,(T_1,\ldots,T_k)}{|DB|}$$

Furthermore, given the conditional independence assumption regarding the labelling process described in an earlier section of this chapter the above expression can be simplified to the following:

$$\forall T_j : j = 1,\ldots,k \; m_{DB}\,(T_1,\ldots,T_k) = \frac{\sum_{i \in DB} \prod_{j=1}^{k} m_{x_j(i)}\,(T_j)}{|DB|}$$

In the case where the dimension k is high then it can be problematic to determine m_{DB} directly from DB. This is due to the 'curse of dimensionality' [8] according to which the amount of data needed to accurately determine a joint probability distribution increases exponentially with the dimension of that distribution. One way of overcoming this problem is to make independence assumptions where possible. This leads to a semi-independent mass relation as follows:

DEFINITION 76 *(Semi-Independent Mass Relation)*
Let $S_1, \ldots, S_w$ be a partition of the variables $x_1, \ldots, x_k$, where $w \leq k$ and w.l.o.g let $S_i = \{x_1, \ldots, x_v\}$ for some $i \in \{1, \ldots, w\}$, then the mass relation describing DB on the basis of partition $S_1, \ldots, S_w$ is given by:

$$\forall T_j \subseteq LA_j : j = 1,\ldots,n \; m_{DB}(T_1,\ldots,T_n) = \prod_{i=1}^{w} m_i(T_j : x_j \in S_i)$$

where m_i is the marginal mass relation over the labels describing S_i such that:

$$\forall T_j \subseteq LA_j : j = 1,\ldots,v \; m_i(T_1,\ldots,T_v) = \frac{1}{|DB|} \sum_{i \in DB} \prod_{j=1}^{v} m_{x_j(i)}(T_j)$$

Here variable dependency is encoded within variable groupings $S_i : i = 1, \ldots, w$ and independence is assumed between the groups. The semi-independence assumption can be strengthened to allow for a fully independent mass relation as follows:

DEFINITION 77 *(The Independent Mass Relation)*
The independent mass relation is the mass relation representing DB based on partition $S_1, \ldots, S_k$ where $S_i = \{x_i\} : i = 1, \ldots, k$

EXAMPLE 78 *Consider a data problem with three attributes x, y and z all into $[0, 30]$. Suppose that all three attributes are labelled with words $\{small(s), medium(m), large(l)\}$ defined according to the trapezoidal appropriateness degrees shown in figure 4.9. Let DB be as is shown in the table from figure 4.10 and assume attribute groupings $S_1 = \{x, y\}$ and $S_2 = \{z\}$.*

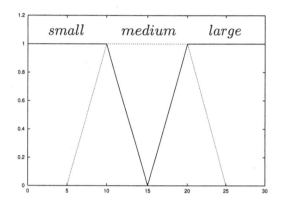

Figure 4.9: Appropriateness degrees for *small*, *medium* and *large*

We can now generate a mass assignment version of this database by replacing x, y and z with $\mathcal{D}_x, \mathcal{D}_y$ and $\mathcal{D}_z$ (see figure 4.11). From this we can combine mass relations for $\mathcal{D}_x \times \mathcal{D}_y$ and $\mathcal{D}_z$ given the prescribed variable grouping to obtain marginal mass relations m_1 and m_2 respectively. Using the semi-independence assumption (definition 76) we can evaluate m_{DB} by taking the product of m_1 and m_2. For example,

$$m_{DB}(\{s, m\}, \{m\}, \{l\}) = m_1(\{s, m\}, \{m\}) \times m_2(\{l\})$$
$$= 0.112 \times 0.52 = 0.05824$$

Given m_{DB} You can then evaluate how appropriate on average a multi-dimensional label expression is as a description of the members of DB. More formally:

$$\forall \theta \in MLE^{(k)} \; \mu_\theta(DB) = \sum_{\langle T_1, \ldots T_k \rangle \in \lambda^{(k)}(\theta)} m_{DB}(T_1, \ldots, T_k)$$

The identification of which attribute groups $S_1, \ldots, S_w$ can be used in practice to model a database DB is clearly a non-trivial problem. It is conceivable that

index	x	y	z
1	5	15	25
2	7	14	28
3	10	10	21
4	20	21	18
5	12	18	17

Figure 4.10: Tableau showing the database DB

index	$\mathcal{D}_x$	$\mathcal{D}_y$	$\mathcal{D}_x \times \mathcal{D}_y$	$\mathcal{D}_z$
1	$\{s\} : 1$	$\{m\} : 1$	$\langle\{s\}, \{m\}\rangle : 1$	$\{l\} : 1$
2	$\{s, m\} : 0.4,$ $\{s\} : 0.6$	$\{s, m\} : 0.2,$ $\{m\} : 0.8$	$\langle\{s, m\}, \{m\}\rangle : 0.32,$ $\langle\{s, m\}, \{s, m\}\rangle : 0.08,$ $\langle\{s\}, \{s, m\}\rangle : 0.12,$ $\langle\{s\}, \{m\}\rangle : 0.48$	$\{l\} : 1$
3	$\{s, m\} : 1$	$\{s, m\} : 1$	$\langle\{s, m\}, \{s, m\}\rangle : 1$	$\{m, l\} : 0.6$ $\{m\} : 0.4$
4	$\{m, l\} : 1$	$\{m, l\} : 0.8$ $\{l\} : 0.2$	$\langle\{m, l\}, \{m, l\}\rangle : 0.8$ $\langle\{m, l\}, \{l\}\rangle : 0.2$	$\{m, l\} : 0.6,$ $\{m\} : 0.4$
5	$\{s, m\} : 0.6$ $\{m\} : 0.4$	$\{m, l\} : 0.6$ $\{m\} : 0.4$	$\langle\{s, m\}, \{m, l\}\rangle : 0.36,$ $\langle\{s, m\}, \{m\}\rangle : 0.24,$ $\langle\{m\}, \{m, l\}\rangle : 0.24,$ $\langle\{m\}, \{m\}\rangle : 0.16$	$\{m, l\} : 0.4,$ $\{l\} : 0.6$
Combined Mass Assignments			$m_1 =$ $\langle\{s\}, \{m\}\rangle : 0.296,$ $\langle\{s, m\}, \{m\}\rangle : 0.112,$ $\langle\{s, m\}, \{s, m\}\rangle : 0.216,$ $\langle\{s\}, \{s, m\}\rangle : 0.024,$ $\langle\{m, l\}, \{m, l\}\rangle : 0.16,$ $\langle\{m, l\}, \{l\}\rangle : 0.04,$ $\langle\{s, m\}, \{m, l\}\rangle : 0.072,$ $\langle\{m\}, \{m, l\}\rangle : 0.048$ $\langle\{m\}, \{m\}\rangle : 0.032$	$m_2 =$ $\{l\} : 0.52$ $\{m, l\} : 0.32$ $\{m\} : 0.16$

Figure 4.11: Mass assignment translation of DB

such groupings might be determined on the basis of prior knowledge relating to dependencies between attributes but such information may not always be available and even if it is available it may be incomplete. In chapter 5 we will investigate learning argorithms that automatically infer groupings from DB in those cases where DB encodes certain input-output relationships as required for a particular superivised learning task.

Summary

In this chapter we have shown how appropriateness measures can be extended to the case where the elements of Ω are described in terms of multiple attributes. The resulting calculus is based on extended definitions for λ-sets and mass assignments, referred to as mass relations in the multi-dimensional case. We have argued for the conditional independence assumption that once the value of a attribute is known, then Your decision regarding the appropriateness of labels for describing that attribute does not dependent on the values of other attributes. This assumption helps to maintain computational tractability and also leads to a number of intuitive properties as identified in theorems 72 and 73.

In addition, to the multi-attribute case we have also extended appropriateness measures and mass assignments to model the description of a set of objects or instances. Given a set of instances $DB \subseteq \Omega$ and a subset of labels T we can infer a mass assignment $m_{DB}(T)$ quantifying the likelihood of selecting an element of DB for which T corresponds to the set labels You judge as appropriate to describe the element. This mass assignment then defines a measure of appropriateness for an expression as a description of DB. In the case that the elements of DB have multiple attributes then the estimation of a mass relation can be problematic due to the curse of dimensionality. This can in part be overcome by a semi-independence assumption whereby the attributes are partitioned into dependent groups. Mass relations are then determined from DB for the joint spaces identified by each attribute grouping and independence is assumed between groups.

Notes

1   $\lambda(\theta) \subseteq LA_j$ refers to the one dimensional λ-set as given in definition 25

Chapter 5

INFORMATION FROM VAGUE CONCEPTS

Let us assume that Your goal as an intelligent agent when You make an assertion is to be both truthful and informative. In this case, we can only assume that the use of vague concepts in such assertions provides You with a flexible tool for conveying information. Furthermore, within the context of our current investigation it is clear that vague concepts should provide information regarding the underlying universe Ω. But what is the nature of this information and how can it best be modelled? For example, if it is know that 'Bill is tall' exactly what does this tell us about Bill's height? In this chapter we will investigate two possible models of the information provided by vague concepts, both resulting from the work of Zadeh ([111],[115]), and investigate their compatibility. We will also propose a framework based on label semantics and relate this to Zadeh's work.

As well as considering the question of what constraints are imposed on the underlying universe Ω by assertions, we shall also consider what information they convey regarding other expressions. This will lead us to investigate a variety of proposed measures for conditionally matching expressions and to assess these with regards to a number of intuitive properties.

Finally, we shall investigate the case where, in label semantics, the information provided is not an expression but rather a mass assignment on label sets. Conditioning on the basis of such information will prove to be particularly relevant when we consider learning linguistic models from data in chapter 5.

5.1 Possibility Theory

In [115] Zadeh proposes that Your assertion of vague concept θ defines a possibility distribution on Ω. From a mathematical perspective a possibility distribution is simply a function $\pi : \Omega \rightarrow [0,1]$ normalized such that $\sup \{\pi(x) : x \in \Omega\} = 1$. Zadeh further suggests that $\pi(x)$ exactly corre-

x	1	2	3	4	5	6	7	8
$\pi(x)$	1	1	1	1	0.8	0.6	0.4	0.2
$P(x)$	0.1	0.8	0.1	0	0	0	0	0

Figure 5.1: Tableau showing possibility and probability distributions for the Hans egg example

sponds to the membership function of θ (i.e. that $\pi(x) = \chi_\theta(x)$). Given this equivalence what remains somewhat unclear is the exact epistemic nature of the constraint that the possibility distribution π places on Ω. Zadeh suggests that $\pi(x)$ measures the possibility of x interpreted as the degree of ease with which the value x can be obtained. Zadeh illustrates this idea using the now famous 'Hans eats eggs' example as follows [115]: Suppose the number of eggs that Hans could conceivably eat for breakfast is contained in the universe $\Omega = \{1, 2, 3, 4, 5, 6, 7, 8\}$. In this context a probability distribution on Ω would quantify the likelihood that Hans will eat a particular number of eggs. For instance, he may be most likely to eat two eggs (probability 0.8), with some small likelihood of him eating either one or three eggs (both probability 0.1) and with there being no chance of him eating more than three eggs (probability 0). On the other hand, according to Zadeh, a possibility distribution on Ω would quantify the ease with which, hypothetically, Hans could eat a particular number of eggs. From this viewpoint, it may be completely possible for Hans to eat up to four eggs (with possibility value 1) and then for him to have increasing difficulty (and decreasing possibility) as the number of eggs increase. The table in figure 5.1 gives a complete possibility and probability distribution for Ω along the lines described above.

For the Hans egg example the concept of difficulty, and hence that of possibility, underlying Zadeh's proposed interpretation would seem to have a clear meaning. We can image Hans increasingly struggling to eat more and more eggs! One wonders, however, whether or not that this clarity of meaning is not largely due to the physical nature of this specific problem. For example, in the case of a possibility distribution on heights induced by the assertion 'Bill is tall' it would seem to make little sense to talk about the difficulty for Bill to achieve a particular height. An alternative, and perhaps more natural, interpretation of possibility distribution arises as a result of its links to Dempster-Shafer theory [95].

In [115] it is proposed that a possibility distribution π on Ω can be extended to a possibility measure Π on subsets of Ω according to:

$$\forall A \subseteq \Omega \; \Pi(A) = \sup\{\pi(x) : x \in A\}$$

In the case where Ω is finite then it can easily be seen that Π corresponds to a plausibility measure where the underlying mass assignment on 2^Ω is consonant (nested) (see chapter 2, definition 14). Furthermore, the possibility distribution π corresponds to the fixed point coverage function of this consonant mass assignment (see chapter 2). From this perspective possibility distributions would appear to encode an uncertain constraint on the values of Ω where the uncertainty is restricted only to the degree of specificity of the constraint. For example, suppose that the universe $\Omega = \{1, 2, 3, 4, 5, 6\}$ corresponds to the values on a dice. A possibility distribution where $\pi\,(1) = 1$, $\pi\,(2) = 0.8$, $\pi\,(3) = 0.6$ and $\pi\,(x) = 0$ for $x \geq 4$ has the following mass assignment:

$$m\,(\{1, 2, 3\}) = 0.6, \; m\,(\{1, 2\}) = 0.2, \; m\,(\{1\}) = 0.2$$

From this we know that either the score on the dice is constrained to being in the set $\{1, 2, 3\}$ with probability 0.6, or to being in the set $\{1, 2\}$ with probability 0.2, or is known to be 1 with probability 0.2. In general, we might perhaps obtain such a mass assignment by asking a population of voters to identify the constraint on the score imposed by the vague concept 'low value'. This naturally links possibility theory with the voting model for fuzzy sets as described in chapter 2.

An alternative justification for the claim that the information regarding Ω provided by vague assertions is possibilistic in nature was proposed by Walley and de Cooman [104] and utilises Walley's theory of imprecise probabilities [102].

5.1.1 An Imprecise Probability Interpretation of Possibility Theory

Walley and de Cooman [104] propose an interpretation of the information provided by linguistic expressions which is related to both prototype semantics and Giles' risk semantics [35] for fuzzy membership functions as described in chapter 2. Specifically, the former claim to have provided a justification for the use of possibility distributions to model the information regarding a variable x contained in assertions of the form 'x is θ' where θ is a vague concept. If we then identify such a possibility distribution with the membership function for θ, as suggested by Zadeh [115], then by implication we have a semantics for fuzzy memberships. However, Walley and de Cooman leave the existence of such a link as an open question.

The interpretation given in [104] uses the theory of imprecise probability as introduced by Walley [103] where probabilistic information takes the form of sets of probability measure characterised by upper and lower probability values, rather than single precise probability measures. More formally, for every subset $S \subseteq \Omega$ we identify an upper and lower probability, $\overline{P}\,(S)$ and $\underline{P}\,(S)$, related according to $\underline{P}\,(S) = 1 - \overline{P}\,(S^c)$, for which we know that the probability of

$S, P(S)$, satisfies:

$$\underline{P}(S) \leq P(S) \leq \overline{P}(S)$$

Upper probability values can also be given an operational interpretation in terms of betting rates suggesting a link between this semantics and that of Giles [35]. Essentially, a person's assessment of $\overline{P}(S)$ is interpreted as a commitment to bet against S at any betting rate greater than $\overline{P}(S)$. See Walley [103] for an in depth discussion of this interpretation.

Initially two types of vague concept, with Ω corresponding to the real numbers, are identified for study. *Increasing predicates* are defined such that $\forall x_1, x_2 \in \Omega$ x_1 satisfies θ at least as well as x_2 if and only if $x_1 \geq x_2$. Similarly *decreasing predicates* are defined such that $\forall x_1, x_2 \in \Omega$ x_1 satisfies θ at least as well as x_2 if and only if $x_1 \leq x_2$. An example of an increasing predicate is the concept tall where Ω corresponds to the set of heights, whereas an example of a decreasing predicate is the concept young with Ω corresponding to the set of ages. Now for increasing predicates Walley and de Cooman claim that learning 'x is θ' provides some evidence against small values of x while not really providing any positive evidence for any particular x value. If we are told that 'John is old' then this would seem to provide evidence against John being less then 20 while not providing any positive evidence has to his actual age. This would suggest that, for increasing predicates, learning 'x is θ' may enable us to determine upper probabilities of the form $\overline{P}(x \leq a)$ for $a \in \Omega$. In effect then we have an upper cumulative probability function defined by $\overline{F}(a) = \overline{P}(x \leq a)$. In general, however, we will want to make inferences regarding the upper probability $\overline{P}(S)$ of any set $S \subseteq \Omega$. Hence, we require a method for extending the upper commutative function $\overline{F}$ on Ω to an upper probability measure on 2^{Ω}. Walley and de Cooman then suggest that such an extension should corresponding to picking the greatest (i.e. least committal) upper probability measure consistent with the assessments $\overline{P}(x \leq a)$ for all $a \in \Omega$. In this case since, given that θ is an increase predicate, we may assume the $F(a)$ is an increasing function on Ω it follows that $\overline{P}(S)$ is determined according to:

$$\overline{P}(S) = \sup\left\{\overline{F}(a) : a \in S\right\}$$

Clearly then in this case $\overline{P}$ is the possibility measure on 2^{Ω} generated by the possibility distribution $\overline{F}$ and hence, for increasing predicates, some justification has been provided for Zadeh's claim that linguistic information leads to possibilistic uncertainty. Walley and de Cooman then go on to suggest that this property can be extended to the so-called monotonic class of predicate that include both increasing and decreasing predicates. Formally, θ is a *monotonic predicate* if there is some measurement function g_θ from Ω into the real numbers such that $\forall x_1, x_2 \in \Omega$ x_1 satisfies θ at least as well as x_2 if and only if

$g_\theta(x_1) \geq g_\theta(x_2)$. Clearly, both increasing and decreasing predicates are special cases of monotonic predicates since for the former we can take $g_\theta(x) = x$ and for the latter $g_\theta(x) = -x$. Intuitively, the idea is that monotonic predicates correspond to those for which there is some notion of θ-ness that naturally defines a total ordering $\succeq_\theta$ on Ω and which can be captured by the real-valued measure g_θ. Formally, the ordering $\succeq_\theta$ can then be defined by $\forall x_1, x_2 \in \Omega$ $x_1 \succeq_\theta x_2$ if and only if $g_\theta(x_1) \geq g_\theta(x_2)$. For example, consider the concept tall defined this time across a finite set of people Ω, the $\succeq_{tall}$ would correspond to the ordering 'taller than' and a natural candidate for g_{tall} is each persons height. Now given this definition we would naturally view the assertion 'x is θ' as providing evidence against elements $y \in \Omega$ occurring low down in the ordering $\succeq_\theta$. In terms of the mapping g_θ this would suggest allocating an upper probability value to sets of the form $\{x \in \Omega : g_\theta(x) \leq a\}$ where $a = g_\theta(y)$ for $y \in \Omega$, thus generating an upper cumulative distribution function $\overline{F}(a) = \overline{P}(\{x \in \Omega : g_\theta(x) \leq a\})$. From this we can use essentially the same argument as in the case of increasing predicates to justify the assumption that $\overline{F}$ is a possibility distribution.

There is an interesting link between this imprecise probabilities semantics and prototype semantics afforded by the identification of a particular class of monotonic predicates. Suppose that for a particular vague concept θ there a number of prototypical cases $\mathcal{P}_\theta$ which definitely satisfy that concept. As before it is then assumed that there is a similarity measure S defined between elements of Ω. From this we can show that θ is a monotonic predicate by defining g_θ as follows:

$$\forall x \in \Omega \; g_\theta(x) = sup\{S(x,y) : y \in \mathcal{P}_\theta\}$$

The imprecise probability semantics of Walley and de Cooman is in many ways an attractive approach to modelling the information conveyed by imprecise concepts, particularly in that it avoids the Bayesian assumption of an uninformative prior. However, it does appear to have a number of drawbacks. One of these regards the ease with which we can identify monotonic predicates. For instance, even if predicate θ intuitively defines a degree of θ-ness and hence an associated ordering $\succeq_\theta$ on Ω, it may be hard to identify the function g_θ in order to explicitly define this ordering. In fact Walley and de Cooman themselves point out this difficulty noting that for the predicate blonde defined on a finite set of people, it is hard to imagine the exact nature of the function g_{blonde} even though we intuitively perceive that some people are blonder than others. However, the problem seems even more widespread since one suspects that even the use of so-called increasing and decreasing predicates is more complex than suggested in [104]. For example, it appears debatable that the concept young is really decreasing at all! Indeed the assertion 'x is young' would intuitively seem to provide some positive evidence against very low ages since if such ages were intended as possibilities then the individual in question would have asserted 'x is very young' or perhaps even 'x is very very young'. From this we make two observations. Firstly, it would seem that the monotonicity or otherwise of a predicate cannot be decided in isolation but only with reference to other predicates in the language. Secondly, even for predicates such as young or tall, in order to define the mapping g_θ we are likely to have to adopt the prototype method. This of course requires the identification of a set of prototypes $\mathcal{P}_\theta$ and a similarity measure S, the difficulty associated with which has already been discussed in chapter 2.

A second drawback of this semantics, significant in the context of this volume, stems from the fact that it is principally a model of the information conveyed by single imprecise concepts and does not consider the relationship between the information provided by a compound expression and that provided by its corresponding component expressions. For example, what is the relationship between the information conveyed by the single assertion 'x is tall and medium' and that conveyed by the two assertions 'x is tall' and 'x is medium'? Such questions are fundamental in the development of a reasoning framework for vague concepts.

5.2   The Probability of Fuzzy Sets

In the case where You have some prior probability distribution on the universe Ω it would seem to be more natural to view a vague concept as providing additional information according to which You may be able to conditionally update this prior. In the context of fuzzy set theory this requires definitions for the probability and conditional probability of fuzzy sets. In [111] Zadeh proposes the following definition for the probability of a fuzzy set as the expected value of its membership function.

DEFINITION 79 *Probability of a Fuzzy Set*
Assume the universe Ω is finite. Let θ be a vague concept defined on universe Ω with membership function χ_θ and let P be a prior probability distribution on Ω then the probability of θ holding is given by:

$$P\left(\theta\right) = \sum_{x \in \Omega} \chi_\theta\left(x\right) P\left(x\right) = E_P\left(\chi_\theta\right)$$

For infinite universes we shall consider the special case where Ω is some closed interval of the real numbers. In this case we assume a prior probability density function f on Ω and the probability of θ is then given by:

$$P\left(\theta\right) = \int_\Omega \chi_\theta\left(x\right) f\left(x\right) dx$$

This definition has a clear justification from the perspective of the random set interpretation of vague concepts as outlined in chapter 2. To see this suppose that we have a vague concept θ, the extension of which corresponds to a random set R_θ with mass assignment m. Now given a probability distribution P on Ω then a natural way to evaluate the probability of θ being true would be to use P to determine the probability of the extension of θ. However, since knowledge of the extension of θ is limited to the mass assignment m over possible subsets of Ω we can, at best, approximate this value. In this case, a natural approximate would be the expected value of the probability of the extension of θ relative to the mass assignment m. This is given by:

$$P(\theta) = \sum_{T \subseteq \Omega} P(T) m(T)$$

Assuming that we take the membership function of θ as corresponding to the fixed point coverage function of the random set R_θ then the above expression is equal to Zadeh's probability value for θ as given in definition 79. This can be seen as follows:

$$\sum_{T \subseteq \Omega} P(T) m(T) = \sum_{T \subseteq \Omega} m(T) \sum_{x \in T} P(x) = \sum_{x \in \Omega} P(x) \sum_{T : x \in T} m(T)$$

$$= \sum_{x \in \Omega} P(x) \chi_\theta(x)$$

Based on the formula for the probability of a fuzzy set (definition 79) Zadeh [111] then proposes the following definition for the posterior distribution on Ω obtained by updating the prior distribution P on the basis of fuzzy information θ:

DEFINITION 80 *Conditional Distributions from Fuzzy Constraints*

$$\forall x \in \Omega \; P(x|\theta) = \frac{\chi_\theta(x) P(x)}{\sum_{x \in \Omega} \chi_\theta(x) P(x)} \; \textit{for the discrete case}$$

$$\forall x \in \Omega \; f(x|\theta) = \frac{\chi_\theta(x) f(x)}{\int_\Omega \chi_\theta(x) f(x) \, dx} \; \textit{for the continuous case}$$

Clearly then in the case of a uniform prior probability distribution $P(x)$, the posterior distribution obtained from the fuzzy constraint θ is simply the normalization of the membership function $\chi_\theta(x)$. Indeed, definition 80 can be viewed as a version of Bayes theorem provided we interpret the membership function $\chi_\theta(x)$ as the likelihood of vague concept θ given value x. This interpretation would seem to be consistent with the semantics for membership functions proposed by Hisdal [44] (discussed in chapter 2).

One aspect of Zadeh's definition 80 that seems particularly unclear is the relationship between the posterior distribution $P(x|\theta)$ and the possibility distribution generated by θ (i.e. $\pi(x) = \chi_\theta(x)$). In [115] it is suggested that possibility distributions should be upper bounds on probability distributions according to which the following 'possibility/probability consistency' principle should hold: For a given possibility distribution π on Ω then a probability distribution P on Ω is consistent with π provided the following inequality holds:

$$\forall x \in \Omega \; P(x) \leq \pi(x)$$

Now since both $P(x|\theta)$ and $\pi(x) = \chi_\theta(x)$ are both derived from the constraint θ then it would seem natural to expect that the former is consistent with the latter. In other words, we would expect that:

$$\forall x \in \Omega \; P(x|\theta) \leq \chi_\theta(x)$$

However, this does not in general hold for definition 80. For a counter example, consider a value $y \in \Omega$ for which $0 < \chi_\theta(y) < 1$ and suppose we have a prior probability distribution such that $P(y) = 1$ and $P(x) = 0$ for all $x \in \Omega$ where $x \neq y$. Then $P(y|\theta) = 1 > \chi_\theta(y)$. Hence, Zadeh's definition of a conditional distribution given a fuzzy constraint would seem to owe more the Bayes' theorem than it does to possibility theory.

An alternative definition for the posterior probability distribution conditional on a fuzzy constraint θ was proposed in series of papers [5], [6], [7] and which is based directly on the Dempster-Shafer interpretation of possibility measures. As with the probability of a vague concept it is argued that $P(x|\theta)$ should correspond to the conditional probability of x given that x is contained in the extension of θ. However, since according to Zadeh [115] Your knowledge of θ is restricted to the possibility distribution $\chi_\theta(x)$ then You are uncertain as to the exact definition of the extension of θ. This uncertainty is represented by the consonant mass assignment m with fixed point coverage function $\chi_\theta(x)$. In this case a natural estimate of $P(x|\theta)$ would be the expected value, relative to mass assignment m, of conditional probabilities $P(x|T)$ across all subsets $T \subseteq \Omega$:

DEFINITION 81 *Conditional Distributions from Mass Assignments*

$$\forall x \in \Omega \; P(x|\theta) = \sum_{T \subseteq \Omega} P(x|T)m(T) = \sum_{T : x \in T} \frac{P(x)}{P(T)}m(T)$$

$$= P(x) \sum_{T : x \in T} \frac{m(T)}{\sum_{y \in T} P(y)}$$

$$\forall x \in \Omega \; f(x|\theta) = f(x) \int_0^{\chi_\theta(x)} \frac{1}{P(\theta_\alpha)} d\alpha \; \text{where } \theta_\alpha = \{x \in \Omega : \chi_\theta(x) \geq \alpha\}$$

Note that θ_α is commonly referred to in the fuzzy logic literature as the α-cut of θ. α-cuts are widely used as a mechanism for extending set functions to fuzzy sets, an idea first proposed by Dubois and Prade in [24].

In the case where the prior $P(x)$ is the uniform distribution then

$$P(x|\theta) = \sum_{T:x \in T} \frac{m(T)}{|T|}$$

This corresponds to the pignistic distribution for mass assignment m as introduced by Smets [97] as a basis for probabilistic decision making within his transferable belief model.

The conditional distribution $P(x|\theta)$ based on mass assignments is consistent with the possibility distribution inferred by fuzzy constraint θ as the following result shows:

THEOREM 82 *If the conditional distribution $P(x|\theta)$ is defined as given in definition 96 then:*

$$\forall x \in \Omega \; P(x|\theta) \le \chi_\theta(x)$$

Proof

$$\forall x \in \Omega \; P(x|\theta) = P(x) \sum_{T:x \in T} \frac{m(T)}{P(T)} = \sum_{T:x \in T} \frac{P(x)}{P(T)} m(T)$$

Clearly this is maximal when $\frac{P(x)}{P(T)} = 1$ for all $T \subseteq \Omega : x \in T$. This occurs when $P(x) = 1$ and $\forall y \in \Omega : y \ne x \; P(y) = 0$. Hence,

$$\sum_{T:x \in T} \frac{P(x)}{P(T)} m(T) \le \sum_{T:x \in T} m(T) = \chi_\theta(x)$$

as required □.

EXAMPLE 83 *Let $\Omega = [0,1]$ and assume that the prior probability distribution is uniform so that $\forall x \in \Omega \; f(x) = 1$. Let θ be a vague concept with a triangular membership function defined by:*

$$\chi_\theta(x) = \begin{cases} 2x & : & x \le 0.5 \\ 2(x-1) & : & x > 0.5 \end{cases}$$

According to Zadeh's definition 80 then

$$f(x|\theta) = \frac{\chi_\theta(x)}{\int_0^1 \chi_\theta(x)\,dx} = \begin{cases} 4x & : & x \le 0.5 \\ 4(x-1) & : & x > 0.5 \end{cases}$$

Alternatively, according to the mass assignment based definition 96 then

$$f\left(x|\theta\right) = \int_0^{\chi_\theta(x)} \frac{1}{P\left(\theta_\alpha\right)d\alpha}\ \textit{where}\ P\left(\theta_\alpha\right)d\alpha = \int_{\theta_\alpha} d\alpha$$

Now for $x < 0.5$ then $\chi_\theta\left(x\right) = 2x$ and $\theta_\alpha = \left[\frac{\alpha}{2}, 1 - \frac{\alpha}{2}\right]$. Hence $P\left(\theta_\alpha\right) = 1 - \alpha$ and

$$f\left(x|\theta\right) = \int_0^{2x} \frac{1}{1-\alpha}d\alpha = -\ln\left(1 - 2x\right)$$

Similarly, if $x > 0.5$ then $f\left(x|\theta\right) = -\ln\left(2x - 1\right)$. In summary (see figure 5.2),

$$f\left(x|\theta\right) = \begin{cases} -\ln\left(1 - 2x\right) & : \quad x < 0.5 \\ -\ln\left(2x - 1\right) & : \quad x > 0.5 \end{cases}$$

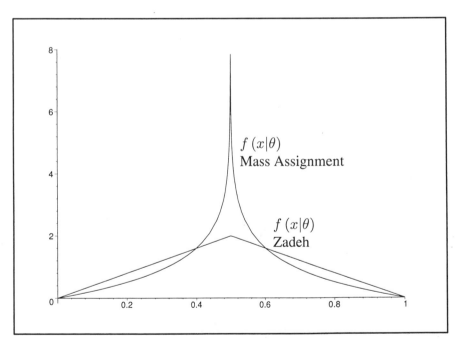

Figure 5.2: Alternative definitions of the conditional distribution $f\left(x|\theta\right)$

5.3   Bayesian Conditioning in Label Semantics

In this section we investigate a possible Bayesian model of the information conveyed by an expression θ from the perspective of label semantics. In this case we assume that when You learn θ You interpret it as meaning that θ is appropriate to describe some unknown value $y \in \Omega$ (i.e. that $\mathcal{D}_y \in \lambda(\theta)$). Hence, given a prior distribution $P(x)$ on Ω we can use Bayes theorem to obtain a posterior distribution conditional on θ as follows:

$$\forall x \in \Omega \; P(x|\theta) = P(y = x|\mathcal{D}_y \in \lambda(\theta))$$
$$= \frac{Pr(\mathcal{D}_y \in \lambda(\theta) | y = x) P(y = x)}{\sum_{x \in \Omega} P(\mathcal{D}_y \in \lambda(\theta) | y = x) P(y = x)}$$

Now

$$P(\mathcal{D}_y \in \lambda(\theta) | y = x) = P(\mathcal{D}_x \in \lambda(\theta)) = \sum_{T \in \lambda(\theta)} m_x(T) = \mu_\theta(x)$$

Hence,

$$\forall x \in \Omega \; P(x|\theta) = \frac{\mu_\theta(x) P(x)}{\sum_{x \in \Omega} \mu_\theta(x) P(x)}$$

Similarly for the continuous case, given prior density f, the corresponding posterior density is:

$$\forall x \in \Omega \; f(x|\theta) = \frac{\mu_\theta(x) f(x)}{\int_\Omega \mu_\theta(x) f(x) \, dx}$$

Clearly then this Bayesian interpretation of the information obtained from expression θ within label semantics is very similar to Zadeh's definition 80 but where the fuzzy set membership function $\chi_\theta(x)$ is replaced by the appropriateness measure $\mu_\theta(x)$. Notice that the denominator in the above expressions corresponds to the probability that some unknown value y can be appropriately described by expression θ. Effectively this is the probability of θ based on appropriateness measure μ_θ as summarised by the following definition:

DEFINITION 84 *Probability of expressions based on Appropriateness Measures*

$$\forall \theta \in LE \; P(\theta) = P(\mathcal{D}_y \in \lambda(\theta)) =$$
$$\sum_{x \in \Omega} P(\mathcal{D}_y \in \lambda(\theta) | y = x) P(x) = \sum_{x \in \Omega} \mu_\theta(x) P(x) = E_P(\mu_\theta)$$

where μ_θ is the appropriateness measure of θ

Clearly this is equivalent to Zadeh's definition of the probability of fuzzy sets (definition 79) but where truth-functional memberships are replaced by functional appropriateness measures. This subtle difference will have important consequences when we consider conditional probabilities for matching vague concepts in a subsequent section of this chapter.

We now illustrate how such Bayesian conditioning can be used to evaluate output values of a system given specific input values when the relationships between input and output variables are described in terms of linguistic expressions. Initially, however, we observe that the conditional distributions defined above can easily be extended to the multi-dimensional case. For example, for the continuous case where we have a joint space $\Omega_1 \times \ldots \times \Omega_k$ and a prior density function $f(x_1, \ldots, x_k)$ then given a multi-dimensional expression θ the posterior density is:

$$\forall \vec{x} \in \Omega_1 \times \ldots \times \Omega_k \ f(x_1, \ldots, x_k|\theta) =$$

$$\frac{\mu_\theta^{(k)}(x_1, \ldots, x_k) f(x_1, \ldots, x_k)}{\int_{\Omega_1} \cdots \int_{\Omega_k} \mu_\theta^{(k)}(x_1, \ldots, x_k) f(x_1, \ldots, x_k) \, d\vec{x}}$$

EXAMPLE 85 *Recall the problem described in example 74 where our knowledge of the relationship between variables x_1 and x_2 corresponds to:*

$$\theta \equiv (medium_1 \wedge \neg large_1 \rightarrow medium_2) \wedge (large_1 \rightarrow small_2)$$

Assuming a uniform prior distribution on $[0, 10]^2$ then the posterior distribution on $\Omega_1 \times \Omega_2$ is given by:

$$\forall x_1 \in \Omega_1, \ x_2 \in \Omega_2 \ f(x_1, x_2|\theta) =$$

$$\frac{\mu_{(m_1 \wedge \neg l_1 \rightarrow m_2) \wedge (l_1 \rightarrow s_2)}^{(2)}(x_1, x_2)}{\int_0^{10}, \int_0^{10} \mu_{(m_1 \wedge \neg l_1 \rightarrow m_2) \wedge (l_1 \rightarrow s_2)}^{(2)}(x_1, x_2) dx_1 dx_2}$$

Now suppose we are given the value of x_1 and we want to calculate the probability of different values of x_2 given this information. In this case we need to evaluate the following conditional distribution:

$$f(x_2|\theta, x_1) = \frac{f(x_1, x_2|\theta)}{f(x_1|\theta)} = \frac{\mu_{(m_1 \wedge \neg l_1 \rightarrow m_2) \wedge (l_1 \rightarrow s_2)}^{(2)}(x_1, x_2)}{\int_0^{10} \mu_{(m_1 \wedge \neg l_1 \rightarrow m_2) \wedge (l_1 \rightarrow s_2)}^{(2)}(x_1, x_2) dx_2}$$

A plot of this distribution as both x_1 and x_2 vary is given in figure 5.3. In the case that $x_1 = 6.5$ the conditional density $f(x_2|\theta, 6.5)$ is shown in figure 5.4. Therefore, in order to obtain an estimate of output x_2 given input $x_1 = 6.5$ we can evaluate the expected value of this distribution:

$$\hat{x_1} = \int_0^{10} x_2 \ f(x_2|\theta, 6.5) dx_2 = 4.5079$$

5.4   Possibilistic Conditioning in Label Semantics

An alternative label semantics model of the information provided by linguistic constraints of the form 'x is θ' can be obtained by considering the possible

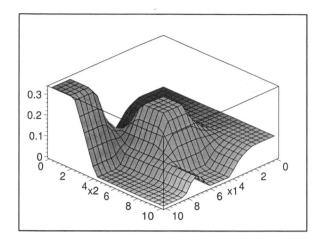

Figure 5.3: Plot of the conditional density $f(x_2|\theta, x_1)$

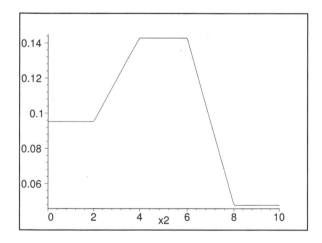

Figure 5.4: Plot of the conditional density $f(x_2|\theta, 6.5)$

values of x for which another agent would actually choose to assert the expression θ. In other words, assuming that another agent asserts that 'x is θ' to You, then You could attempt to identify the subset $X(\theta) \subseteq \Omega$ of possible values for x that would result in such an assertion. Implicitly, in this case we would need to assume that there is a shared interpretation of the meaning of the labels between You and the other agent (represented in terms of appropriateness measures) and also that You have some knowledge of the underlying process according to which assertions are made within the relevant shared linguistic context. For example, You might assume that an agent would only assert 'x is θ' if the measure of appropriateness of θ for describing x exceeds some threshold α. In this case the set of possible values for x is also a function of α given by:

$$\forall \alpha \in [0,1] \ X(\theta, \alpha) = \{x \in \Omega : \mu_\theta(x) \geq \alpha\} = \theta_\alpha$$

[1] Clearly then Your uncertainty regarding $X(\theta, \alpha)$ is directly dependent on Your uncertainty regarding the threshold α that the other agent is likely to have used to select possible assertions. Suppose that You identify a probability distribution on $[0,1]$ with a density function f to represent this uncertainty. You could then identify a measure of the possibility of an element $x \in \Omega$ given the assertion θ as the probability that the value α is such that $x \in X(\theta, \alpha)$:

$$\forall x \in \Omega \ \pi(x) = P(\{\alpha \in [0,1] : x \in X(\theta, \alpha)\}) = P(\{\alpha \in [0,1] : x \in \theta_\alpha\})$$

$$= \int_0^{\mu_\theta(x)} f(\alpha) \, d\alpha = F(\mu_\theta(x))$$

where $F(\alpha)$ is the cumulative distribution function generated by density f. In the case where f is the uniform distribution then:

$$\pi(x) = \mu_\theta(x)$$

Hence, in this case we would be associating the possibility distribution generated by expression with the appropriateness measure of θ in the same way as Zadeh [115] associated possibility with membership. However, there would seem to be little justification for You to take f to be the uniform distribution since assuming that the other agent is trying to be truthful it is unlikely that they would pick a threshold value less than 0.5. This would suggest that f should be skewed towards the interval $[0.5, 1]$. For example, You might take f to be uniform on $[0.5, 1]$ and zero on $[0, 0.5)$. In this case:

$$\pi(x) = \begin{cases} 0 & : \ \mu_\theta(x) \in [0, 0.5) \\ 2\mu_\theta(x) - 1 & : \ \mu_\theta(x) \in [0.5, 1] \end{cases}$$

More likely, however, You would take f to be symmetric about Your own choice of α parameter value for making assertions based on the view that Your fellow agents are similar enough to You to warrant such an assumption.

When considering the information provided by an assertion θ there would seem to be additional information contained in the fact no other more specific assertion was made. In the model for assertions proposed in chapter 3 it was suggested that an agent might aim to identify the most specific expression fulfilling the appropriateness threshold α. Hence, from the assertion θ You can infer that in addition to $\mu_\theta(x) \geq \alpha$ it must also hold that for all more specific expressions φ, $\mu_\varphi(x) < \alpha$. In this case:

$$X(\theta, \alpha) = \{x \in \Omega : \mu_\theta(x) \geq \alpha, \ \forall \varphi \models \theta \ \mu_\varphi(x) < \alpha\} = \theta_\alpha \cap \left(\bigcup_{\varphi : \varphi \models \theta} \varphi_\alpha \right)^c$$

Hence,

$$\pi\left(x\right) = P\left(\left\{\alpha \in [0,1] : x \in X\left(\theta, \alpha\right)\right\}\right) = \int_{\theta_{\alpha} \cap \left(\bigcup_{\varphi:\varphi\models\theta} \varphi_{\alpha}\right)^{c}} f\left(\alpha\right) d\alpha$$

EXAMPLE 86 *Consider the labels young, middle aged and old for the attribute age with appropriateness measures defined as in example 75. Supposing another agent tells You that Bill is middle aged (i.e. that age is middle aged) then You could infer the following possibility distributions on the age of Bill: Taking* $X\left(ma, \alpha\right) = \left\{x \in \Omega : \mu_{ma}\left(x\right) \geq \alpha\right\}$ *and assuming a uniform distribution for* α *values across* $[0,1]$ *then* $\forall x \in \Omega$ $\pi\left(x\right) = \mu_{\theta}\left(x\right)$ *corresponding to the solid black line in figure 5.5. If* $X\left(ma, \alpha\right)$ *remains unchanged but it is assumed that* α *can only take values greater than or equal to 0.5 and furthermore that there is a uniform distribution across such values then:*

$$\pi\left(x\right) = \begin{cases} 0 & : & \mu_{ma}\left(x\right) \in [0, 0.5) \\ 2\mu_{ma}\left(x\right) - 1 & : & \mu_{ma}\left(x\right) \in [0.5, 1] \end{cases}$$

This possibility distribution is shown by the dashed line in figure 5.5. In the

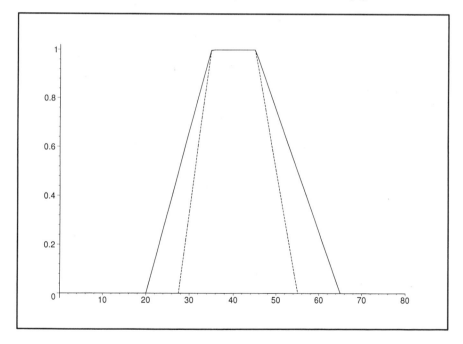

Figure 5.5: Possibility distributions where $\pi\left(x\right) = \mu_{ma}\left(x\right)$ (black line) and $\pi\left(x\right)$ defined such that $\pi\left(x\right) = 0$ if $\mu_{ma}\left(x\right) < 0.5$ and $\pi\left(x\right) = 2\mu_{ma}\left(x\right) - 1$ if $\mu_{ma}\left(x\right) \geq 0.5$ (dashed line).

case where

$$X\left(ma, \alpha\right) = \left\{x \in \Omega : \mu_{ma}\left(x\right) \geq \alpha, \forall\varphi \models middle\ aged\ \mu_{\varphi}\left(x\right) < \alpha\right\}$$

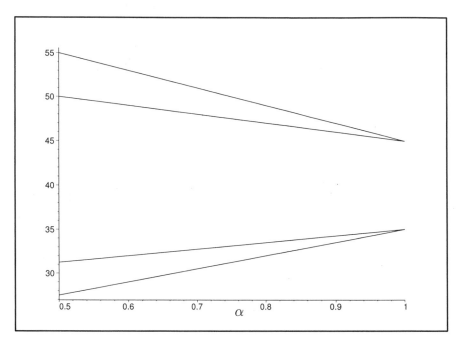

Figure 5.6: $X(ma,\alpha) = \{x \in \Omega : \mu_{ma}(x) \geq \alpha,\ \forall\varphi \models middle\ aged\ \mu_\varphi(x) < \alpha\}$ plotted as a function of α

we need only consider those expressions φ where $\varphi \models middle\ aged$ and for which $\exists x \in \Omega$ such that $\mu_\varphi(x) > 0$. Given the definition of the appropriateness measures for young, middle aged and old as shown in figure 75 then (up to equivalence) there are only three such expressions: young $\wedge$ middle aged, $\neg$young $\wedge$ middle aged $\wedge$ $\neg$old and middle aged $\wedge$ old. Hence,

$$X(ma,\alpha) = ma_\alpha \cap \left((y \wedge ma)_\alpha \cup (\neg y \wedge ma \wedge \neg o)_\alpha \cup (ma \wedge o)_\alpha\right)^c$$

Now

$$\forall\alpha \in [0,1]\ ma_\alpha = [15\alpha + 20, 65 - 20\alpha]$$

For $\alpha \leq 0.5$

$$(\neg y \wedge ma \wedge \neg o)_\alpha = [27.5 + 7.5\alpha, 15 - 10\alpha],$$
$$(y \wedge ma)_\alpha = [15\alpha + 20, 35 - 15\alpha],$$
$$(ma \wedge o)_\alpha = [45 + 20\alpha, 65 - 20\alpha]$$

Therefore,

$$(\neg y \wedge ma \wedge \neg o)_\alpha \cup (y \wedge ma)_\alpha \cup (ma \wedge o)_\alpha = [15\alpha + 20, 65 - 20\alpha] = ma_\alpha$$

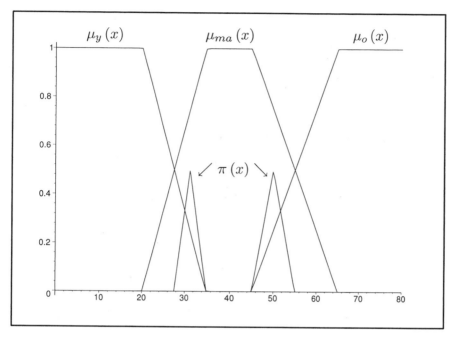

Figure 5.7: Possibility distribution generated from $X\,(ma,\alpha)$ as defined in figure 5.6 and assuming that α is uniformly distributed across $[0.5, 1]$ and has zero probability for values less that 0.5

Hence,

$$\forall \alpha \leq 0.5\; X\,(ma,\alpha) = \emptyset$$

For $\alpha > 0.5$

$$(y \wedge ma)_\alpha = \emptyset, \; (ma \wedge o)_\alpha = \emptyset, \; (\neg y \wedge ma \wedge \neg o)_\alpha = [27.5 + 7.5\alpha, 55 - 10\alpha]$$

Hence,

$$\forall \alpha > 0.5\; X\,(ma,\alpha) = [15\alpha + 20, 65 - 20\alpha] - [27.5 + 7.5\alpha, 55 - 10\alpha]$$
$$= [15\alpha + 20, 27.5 + 7.5) \cup (55 - 10\alpha, 65 - 20\alpha]$$

A plot of $X\,(ma,\alpha)$ for α between 0.5 and 1 is shown in figure 5.6.
  Now assuming that α can only take values greater than or equal to 0.5 and furthermore that there is a uniform distribution across such values then:

$$\pi\,(x) = P\,(\alpha \in [0.5, 1] : x \in [15\alpha + 20, 27.5 + 7.5) \cup (55 - 10\alpha, 65 - 20\alpha])$$

In this case (see figure 5.7):

$$
\pi(x) = \begin{cases}
\frac{x}{20} - \frac{9}{4} & : \; x \in [45, 50] \\
\frac{11}{4} - \frac{x}{20} & : \; x \in (50, 55] \\
\frac{x}{15} - \frac{11}{6} & : \; x \in [27.5, 31.25] \\
\frac{7}{3} - \frac{x}{15} & : \; x \in [31.25, 35] \\
0 & : \; otherwise
\end{cases}
$$

5.5 Matching Concepts

Suppose it is known that the variable x is constrained by the linguistic expression φ. In this case, what is the degree to which x can also be described by some other expression θ. For example, if we know that 'Bill is *tall*' what is Your level of belief that Bill is also *very tall*? This is an important question that takes on special significance in the area of fuzzy or possibilistic logic programming ([3], [26] and [4]). A recent logic programming implementation of label semantics is described by Noi and Cao [74]. In this context a mechanism is required by which we can evaluate the semantic match (or unification) of an expression θ, forming part of a query, with a given expression φ in the knowledge base.

5.5.1 Conditional Probability and Possibility given Fuzzy Sets

In [111] Zadeh extends the definition of probability of a fuzzy set to conditional probability as follows:

DEFINITION 87 *Conditional Probability of Fuzzy Sets*
For vague concepts $\theta, \varphi \in LE$

$$
P(\theta|\varphi) = \frac{E_P(\chi_\theta(x) \times \chi_\varphi(x))}{E_P(\chi_\varphi(x))}
$$

This means that:

$$
P(\theta|\varphi) = \frac{P(\theta \wedge \varphi)}{P(\varphi)}
$$

where $P(\theta \wedge \varphi)$ and $P(\varphi)$ are as given in definition 79, provided $\chi_{\theta \wedge \varphi}(x) = \chi_\theta(x) \times \chi_\varphi(x)$. In other words, provided that $f_\wedge$ is the product t-norm.

Zadeh [111] motivates the choice of the product t-norm in definition 87 by means of the following example. Suppose that we have attributes x_1 and x_2 with associated universes Ω_1 and Ω_2. Let θ be an expression describing x_1 and φ be an expression describing x_2 so that in terms of the notation introduced in chapter 4, $\theta \in LE_1$, and $\varphi \in LE_2$. In terms of membership functions this means that $\chi_\theta(x_1, x_2) = \chi_\theta(x_1)$ and $\chi_\varphi(x_1, x_2) = \chi_\varphi(x_2)$. Now further suppose that a priori x_1 and x_2 are assumed to be independent so that the joint prior distribution P on $\Omega_1 \times \Omega_2$ is the product of the two marginals P_1 and P_2 on Ω_1 and Ω_2 respectively. In this case, according to Zadeh, the information provided by θ is independent of that provided by φ and hence $P(\theta|\varphi) = P(\theta)$ and $P(\varphi|\theta) = P(\varphi)$. We shall refer to the principle underlying this example as Zadeh's independence principle (ZIP).

Suppose then that we choose to define the conditional probability of θ given φ by extending the classical definition so that:

$$P_t(\theta|\varphi) = \frac{P(\theta \wedge \varphi)}{P(\varphi)}$$

where the probability of vague concepts are evaluated according to Zadeh's definition 79 and the concepts themselves are combined according to a truth-functional calculus based on t-norms and t-conorms as described in chapter 2. The subscript t in $P_t(\bullet|\bullet)$ is included to emphasise the dependency of this type of definition for conditional probability on the choice of t-norm used in the underlying truth-functional calculus for fuzzy sets. In this case ZIP clearly holds when the t-norm $f_\wedge$ is the product and in fact the product is the only t-norm for which it holds. To see this suppose that Ω_1 and Ω_2 are both finite universes and that for some $y_1 \in \Omega_1$ and $y_2 \in \Omega_2$, $P(y_1, y_2) = 1$. Hence, for the independence property to hold then the equation

$$f_\wedge(\chi_\theta(y_1), \chi_\varphi(y_2)) = \chi_\theta(y_1) \times \chi_\varphi(y_2)$$

must hold for all values of $\chi_\theta(y_1)$ and $\chi_\varphi(y_2)$ subject only to the constraint that $\chi_\varphi(y_2) > 0$. Hence, given the boundary condition C1 (chapter 2) for t-norms then:

$$\forall a, b \in [0, 1] \quad f_\wedge(a, b) = a \times b$$

We now introduce two properties that we would argue should be satisfied by any definition of conditional probability for matching vague concepts. Throughout the remainder of this section we shall use these properties together with Zadeh's independence principle, introduced above, to help us assess different definitions of conditional probability for vague concepts.

CPR1:

$$\forall \theta \in LE \; P(\theta|\theta) = 1$$

Intuitively we might justify this property by noting that uncertainty cannot result from certainty even when the concepts involved are vague. If You know with certainty that 'Bill is tall' then You are certain that 'Bill is tall'.

CPR2:

$$\forall \theta, \varphi \in LE \ P(\theta|\varphi) + P(\neg\theta|\varphi) = 1$$

This property states that the underlying additivity of probability should be preserved when it is generalised to vague concepts.

It is trivial to see that CPR1 is not satisfied by Zadeh's definition 87. Indeed for a t-norm based calculus it can easily be seen that for CPR1 to hold then $f_\wedge$ must be idempotent and therefore must correspond to min (see theorem 2). Hence, there is no choice of t-norm for which $P_t(\bullet|\bullet)$ satisfies both ZIP and CPR1. Conditional probability as given in definition 87 does satisfy CPR2, however, and as can be seen from the following theorem [5], does so for continuous t-norms if and only if $f_\wedge$ is the product t-norm.

THEOREM 88 *Assuming that $f_\wedge$ is continuous on $[0,1]$ and (for simplicity) that Ω is finite then*

$$\forall \theta, \varphi \in LE \ P_t(\theta|\varphi) + P_t(\neg\theta|\varphi) = 1$$

if and only if

$$\forall a, b \in [0,1] \ f_\wedge(a,b) = a \times b$$

Proof
($\Leftarrow$) *Follows trivially*
($\Rightarrow$)
For some $y \in \Omega$ let $\chi_\theta(y) = a$, $\chi_\varphi(y) = b$ and $\forall x \neq y \ \chi_\theta(x) = \chi_\varphi(x) = 0$ then allowing a and b to range between 0 and 1 it follows from CPR2 that:

$$(*) \ \forall a, b \in [0,1] \ f_\wedge(a,b) + f_\wedge(1-a,b) = b$$

By $(*)$ we have that:

$$\forall a, b, c \in [0,1] \ f_\wedge(f_\wedge(a,b),c) + f_\wedge(f_\wedge(a,b),1-c) = f_\wedge(a,b)$$

Then by the associativity of $f_\wedge$ it follows that:

$$f_\wedge(a, f_\wedge(b,c)) + f_\wedge(a, f_\wedge(b,1-c)) = f_\wedge(a,b)$$
$$\Rightarrow \ f_\wedge(a, f_\wedge(b,c)) + f_\wedge(a, b - f_\wedge(b,c)) = f_\wedge(a,b) \ \text{by} \ (*)$$

Now since $f_\wedge(b,c) \leq \min(b,c)$ it follows that $f_\wedge(b,0) = 0$. Also by C1 (see chapter 2) we have that $f_\wedge(b,1) = b$. Hence, since $f_\wedge$ is continuous we have by the intermediate value theorem that $\forall r \in [0,b] \ \exists c \in [0,1]$ such that $f_\wedge(b,c) = r$. Therefore,

$$\forall r \in [0,b] \ f_\wedge(a,r) + f_\wedge(a,b-r) = f_\wedge(a,b)$$

Letting $s = b - r$ gives that $\forall s, r \in [0, 1]$ such that $s + r \in [0, 1]$

$$f_{\wedge}(a, r) + f_{\wedge}(a, s) = f_{\wedge}(a, r + s)$$

For constant a, this is Cauchy's equation (see [1]) and by continuity of $f_{\wedge}$ it follows that:

$$f_{\wedge}(a, b) = k(a) b$$

By symmetry of $f_{\wedge}$ (C3) we have that $k(a) b = k(b) a$ and letting $b = 1$ gives $k(a) = k'a$ and therefore $f_{\wedge}(a, b) = k'ab$. Further, since $f_{\wedge}(a, 1) = a$ we have that $k' = 1$ $\square$

Notice that Zadeh's definition of conditional probability for matching vague concepts is closely linked to Zadeh's definition of a conditional distribution given a fuzzy set (definition 80). Specifically, $P(\theta|\varphi)$ corresponds to the expected value of the membership function $\chi_\theta(x)$ relative to the conditional distribution $P(x|\varphi)$ evaluated according to definition 80. That is, for a finite universe Ω, we have that:

$$\forall \theta, \varphi \in LE \; P(\theta|\varphi) = \frac{\sum_{x \in \Omega} \chi_\theta(x) \chi_\varphi(x) P(x)}{\sum_{x \in \Omega} \chi_\varphi(x) P(x)}$$

$$= \sum_{x \in \Omega} \chi_\theta(x) \left(\frac{\chi_\varphi(x) P(x)}{\sum_{x \in \Omega} \chi_\varphi(x) P(x)} \right) = \sum_{x \in \Omega} \chi_\theta(x) P(x|\varphi) = E_{P(\bullet|\varphi)}(\chi_\theta(x))$$

An alternative definition of conditional probability for matching vague concepts was given in [5] and will prove to be related to conditional distributions based on mass assignments as given in definition 96. This conditional measure is most naturally expressed in terms of α-cuts as follows:

DEFINITION 89

$$\forall \theta, \varphi \in LE \; P_I(\theta|\varphi) = \int_0^1 \int_0^1 P(\theta_\alpha|\varphi_\beta) \, d\beta d\alpha$$

For a finite universe Ω this measure can be expressed in terms of consonant mass assignments m_1 and m_2 with fixed point coverage functions $\chi_\theta(x)$ and $\chi_\varphi(x)$ respectively, as follows:

$$P_I(\theta|\varphi) = \sum_{S : S \subseteq \Omega} \sum_{T : T \subseteq \Omega} P(S|T) m_1(S) m_2(T)$$

Here $m_1(S) = \nu(\{\alpha : \theta_\alpha = S\})$ and $m_2(T) = \nu(\{\alpha : \varphi_\alpha = T\})$ where ν is the Lebesgue measure on Borel subsets of $[0, 1]$.

This definition has a rather natural interpretation in terms of the voting model for vague concepts has outlined in chapter 2. In this model $m_1(S)$ and $m_2(T)$ correspond to the proportion of voters who select S as the extension of θ and T as the extension of φ respectively. Conceptually speaking, one approach to estimating $P(\theta|\varphi)$ within the voting model would be to pick two voters at random from the population V and ask the first for the extension of θ and the second for the extension of φ. The value of $P(\theta|\varphi)$ would then be the estimated as the corresponding classical (crisp) conditional probability based on these two extensions. (i.e. if the first voter selects S as the extension of θ and the second voter selects T as the extension of φ then $P(\theta|\varphi)$ would be estimated by $P(S|T)$). If this experiment was then repeated a large number of times and assuming that the two voter were always picked independently then in the limit the expected value of the estimate of $P(\theta|\varphi)$ is given by the measure P_I from definition 89. The subscript I in the notation P_I represents the independence of the choice of voters in the above voting model or alternatively of the α-cut levels for θ and φ in definition 89.

We now show formally that the measure P_I is directly related to the conditional probability distribution given in definition 96

THEOREM 90 *Assuming Ω is finite (for simplicity) then*

$$\forall \theta, \varphi \in LE \; P_I(\theta|\varphi) = \sum_{x \in \Omega} \chi_\theta(x) P(x|\varphi) = E_{P(\bullet|\varphi)}(\chi_\theta(x))$$

where $P(x|\varphi)$ is the conditional distribution derived from the mass assignment m_2 as given in definition 96.

Proof

$$P_I(\theta|\varphi) = \sum_{S \subseteq \Omega} \sum_{T \subseteq \Omega} P(S|T) m_1(S) m_2(T) =$$

$$\sum_{S \subseteq \Omega} \sum_{T \subseteq \Omega} \frac{P(S \cap T)}{P(T)} m_1(S) m_2(T) =$$

$$\sum_{S \subseteq \Omega} m_1(S) \sum_{T \subseteq \Omega} \frac{P(S \cap T)}{P(T)} m_2(T) =$$

$$\sum_{S \subseteq \Omega} m_1(S) \sum_{T \subseteq \Omega} \left(\sum_{x \in S} P(x|T) \right) m_2(T) =$$

$$\sum_{S \subseteq \Omega} m_1(S) \sum_{x \in S} \sum_{T \subseteq \Omega} P(x|T) m_2(T) =$$

$$\sum_{x \in \Omega} \sum_{S:x \in S} m_1(S) P(x) \sum_{T:x \in T} \frac{m_2(T)}{P(T)} = \sum_{x \in \Omega} \chi_\theta(x) P(x|\varphi) \quad \square$$

P_I can be shown to satisfy ZIP as follows: Under the conditions of the ZIP we have that:

$\theta_\alpha = \{\langle x_1, x_2 \rangle : \chi_\theta(x_1) \geq \alpha\} = \{x_1 : \chi_\theta(x_1) \geq \alpha\} \times \Omega_2$ and similarly

$\varphi_\beta = \{\langle x_1, x_2 \rangle : \chi_\varphi(x_2) \geq \beta\} = \Omega_1 \times \{x_2 : \chi_\varphi(x_2) \geq \beta\}$ hence

$\theta_\alpha \cap \varphi_\beta = \{x_1 : \chi_\theta(x_1) \geq \alpha\} \times \{x_2 : \chi_\varphi(x_2) \geq \beta\}$

From this we have that:

$$P_I(\theta|\varphi) = \int_0^1 \int_0^1 \frac{P(\{x_1 : \chi_\theta(x_1) \geq \alpha\} \times \{x_2 : \chi_\varphi(x_2) \geq \beta\})}{P(\Omega_1 \times \{x_2 : \chi_\varphi(x_2) \geq \beta\})} d\beta d\alpha =$$

$$\int_0^1 \int_0^1 \frac{P_1(\{x_1 : \chi_\theta(x_1) \geq \alpha\}) \times P_2(\{x_2 : \chi_\varphi(x_2) \geq \beta\})}{P_2(\{x_2 : \chi_\varphi(x_2) \geq \beta\})} d\beta d\alpha =$$

$$\int_0^1 P_1(\{x_1 : \chi_\theta(x_1) \geq \alpha\}) d\alpha = P(\theta)$$

We now evaluate P_I in terms of properties CPR1 and CPR2. In fact it can easily be seen that P_I does not satisfy CPR1 from the following counter example: Let $\Omega = \{y_1, y_2\}$ and suppose that $\chi_\theta(y_1) = 1$ and $\chi_\theta(y_2) = 0.3$ and that P is the uniform distribution. From this it follows that $m_1(\{y_1, y_2\}) = 0.3$ and $m_1(\{y_1\}) = 0.7$ and hence:

$$P_I(\theta|\theta) = P(\{y_1, y_2\}|\{y_1, y_2\})(0.3)^2 + P(\{y_1, y_2\}|\{y_1\})(0.3)(0.7) +$$

$$P(\{y_1\}|\{y_1\})(0.7)^2 + P(\{y_1\}|\{y_1, y_2\})(0.7)(0.3) =$$

$$1(0.3)^2 + 1(0.3)(0.7) + 1(0.7)^2 + 0.5(0.7)(0.3) = 0.895 \neq 1$$

P_I trivially satisfies CPR2 by theorem 90 since:

$$P_I(\theta|\varphi) + P_I(\neg\theta|\varphi) = \left(\sum_{x\in\Omega} \chi_\theta(x) P(x|\varphi)\right) + \left(\sum_{x\in\Omega} \chi_{\neg\theta}(x) P(x|\varphi)\right) =$$

$$\sum_{x\in\Omega} (\chi_\theta(x) + \chi_{\neg\theta}(x)) P(x|\varphi) = \sum_{x\in\Omega} P(x|\varphi) = 1$$

A variant of the α-cut definition of conditional probability was proposed in [6] and assumes that the same cut level are selected for both θ and φ.

DEFINITION 91

$$\forall \theta \in LE\ P_{dp}(\theta|\varphi) = \int_0^1 P(\theta_\alpha|\varphi_\alpha) d\alpha$$

In terms of the voting model then definition 91 is motivated by the assumption that the same voter, selected at random from population V, is asked to provide both the extension of θ and that of φ. The subscript dp in the notation P_{dp} represents this dependency.

As with P_I it can easily be seen that P_{dp} satisfies ZIP as follows: For θ and φ defined in ZIP we have that:

$$\theta_\alpha = \{x_1 : \chi_\theta(x_1) \geq \alpha\} \times \Omega_1$$
$$\varphi_\alpha = \Omega_2 \times \{x_2 : \chi_\varphi(x_2) \geq \alpha\} \text{ and hence}$$
$$\theta_\alpha \cap \varphi_\alpha = \{x_1 : \chi_\theta(x_1) \geq \alpha\} \times \{x_2 : \chi_\varphi(x_2) \geq \alpha\}$$

From this we have that:

$$P_{dp}(\theta|\varphi) = \int_0^1 \frac{P(\{x_1 : \chi_\theta(x_1) \geq \alpha\} \times \{x_2 : \chi_\varphi(x_2) \geq \alpha\})}{P(\Omega_1 \times \{x_2 : \chi_\varphi(x_2) \geq \alpha\})} d\alpha =$$
$$\int_0^1 \frac{P_1(\{x_1 : \chi_\theta(x_1) \geq \alpha\}) \times P_2(\{x_2 : \chi_\varphi(x_2) \geq \alpha\})}{P_2(\{x_2 : \chi_\varphi(x_2) \geq \alpha\})} d\alpha =$$
$$\int_0^1 P_1(\{x_1 : \chi_\theta(x_1) \geq \alpha\}) d\alpha = P(\theta)$$

Furthermore, it trivially follows that P_{dp} satisfies CPR1 since:

$$P_{dp}(\theta|\theta) = \int_0^1 P(\theta_\alpha|\theta_\alpha) d\alpha = \int_0^1 d\alpha = 1$$

However as can be seen from the following counter example P_{dp} does not in general satisfy CPR2: Let $\Omega = \{y_1, y_2, y_3, y_4\}$ and assume that $\chi_\theta(y_1) = 1$, $\chi_\theta(y_2) = 0.8$, $\chi_\theta(y_3) = 0.6$, $\chi_\theta(y_4) = 0$ and that $\chi_\varphi(y_3) = 1$, $\chi_\varphi(y_2) = 0.9$, $\chi_\varphi(y_1) = 0.2$, $\chi_\varphi(y_4) = 0$. From this we have that $\chi_{\neg\theta}(y_4) = 1$, $\chi_{\neg\theta}(y_3) = 0.4$, $\chi_{\neg\theta}(y_2) = 0.2$, $\chi_{\neg\theta}(y_1) = 0$. Also assume that P is the uniform distribution on Ω. Now evaluating the α-cuts for θ, $\neg\theta$ and φ we have that:

$$\theta_\alpha = \begin{cases} \{y_1, y_2, y_3\} & : \quad \alpha \in [0, 0.6] \\ \{y_1, y_2\} & : \quad \alpha \in (0.6, 0.8] \\ \{y_1\} & : \quad \alpha \in (0.8, 1] \end{cases}$$

$$(\neg\theta)_\alpha = \begin{cases} \{y_2, y_3, y_4\} & : \quad \alpha \in [0, 0.2] \\ \{y_3, y_4\} & : \quad \alpha \in (0.2, 0.4] \\ \{y_4\} & : \quad \alpha \in (0.4, 1] \end{cases}$$

$$\varphi_\alpha = \begin{cases} \{y_1, y_2, y_3\} & : \quad \alpha \in [0, 0.2] \\ \{y_2, y_3\} & : \quad \alpha \in (0.2, 0.9] \\ \{y_3\} & : \quad \alpha \in (0.9, 1] \end{cases}$$

From this we have that:

$$P_{dp}(\theta|\varphi) = \int_0^{0.2} P(\{y_1, y_2, y_3\} \mid \{y_1, y_2, y_3\}) \, d\alpha +$$

$$\int_{0.2}^{0.6} P(\{y_1, y_2, y_3\} \mid \{y_2, y_3\}) \, d\alpha$$

$$+ \int_{0.6}^{0.8} P(\{y_1, y_2\} \mid \{y_2, y_3\}) \, d\alpha + \int_{0.8}^{0.9} P(y_1 \mid \{y_2, y_3\}) \, d\alpha +$$

$$\int_{0.9}^{1} P(\{y_1\} \mid \{y_3\}) \, d\alpha$$

$$= 0.2\,(1) + 0.4\,(1) + 0.2\,(0.5) + 0.1\,(0) + 0.1\,(0) = 0.7$$

Whereas

$$P_{dp}(\neg\theta|\varphi) = \int_0^{0.2} P(\{y_2, y_3, y_4\} \mid \{y_1, y_2, y_3\}) \, d\alpha +$$

$$\int_{0.2}^{0.4} P(\{y_3, y_4\} \mid \{y_2, y_3\}) \, d\alpha$$

$$+ \int_{0.4}^{0.9} P(\{y_4\} \mid \{y_2, y_3\}) \, d\alpha + \int_{0.9}^{1} P(\{y_4\} \mid \{y_3\}) \, d\alpha$$

$$= 0.2\left(\frac{2}{3}\right) + 0.2\left(\frac{1}{2}\right) + 0.5\,(0) + 0.1\,(0) = 0.233333$$

Hence,

$$P_{dp}(\theta|\varphi) + P_{dp}(\neg\theta|\varphi) = 0.9333333 \neq 1$$

5.5.2 Conditional Probability in Label Semantics

In the context of label semantics when evaluating $P(\theta|\varphi)$ we are essentially evaluating the probability that θ is an appropriate description for some unknown value $y \in \Omega$ (i.e. that $\mathcal{D}_y \in \lambda(\theta)$) given the knowledge that φ is an appropriate description for y (i.e. that $\mathcal{D}_y \in \lambda(\varphi)$). This motivates the following definition of the conditional probability for matching vague concepts based on appropriateness measures.

DEFINITION 92

$$\forall \theta, \varphi \in LE \; P_{ls}(\theta|\varphi) = P(\mathcal{D}_y \in \lambda(\theta) \mid \mathcal{D}_y \in \lambda(\varphi))$$

$$= \frac{P(\mathcal{D}_y \in \lambda(\theta \wedge \varphi))}{P(\mathcal{D}_y \in \lambda(\varphi))} = \frac{P(\theta \wedge \varphi)}{P(\varphi)} = \frac{E_P(\mu_{\theta \wedge \varphi}(x))}{E_P(\mu_\varphi(x))}$$

Here $P(\theta \wedge \varphi)$ and $P(\varphi)$ are the probabilities of expressions $\theta \wedge \varphi$ and φ respectively, evaluated according to definition 84. Hence for finite Ω we have that

$$P_{ls}(\theta|\varphi) = \frac{\sum_{x \in \Omega} \mu_{\theta \wedge \varphi}(x) P(x)}{\sum_{x \in \Omega} \mu_{\theta}(x) P(x)}$$

and for a continuous universe we have:

$$P_{ls}(\theta|\varphi) = \frac{\int_{\Omega} \mu_{\theta \wedge \varphi}(x) f(x) \, dx}{\int_{\Omega} \mu_{\theta}(x) f(x) \, dx}$$

Definition 92 can also be expressed in terms of a conditional mass assignment on labels given that x is described by θ. Suppose You learn that 'x is θ' (i.e. that $\mathcal{D}_x \in \lambda(\theta)$) how does this effect Your belief regarding what is the set of labels with which it is appropriate to describe x? To answer this question we begin by considering Your prior belief in $\mathcal{D}_x$. If You have no information regarding x beyond the prior distribution P on Ω then You might well estimate Your beliefs about the set of appropriate labels for x as a prior mass assignment derived from P according to the theorem of total probability as in the following definition:

DEFINITION 93 *Prior Mass Assignment on Labels*

$$\forall T \subseteq LA \; pm(T) = E_P(m_x(T))$$

For finite Ω this corresponds to:

$$pm(T) = \sum_{x \in \Omega} m_x(T) P(x)$$

and in the continuous case:

$$pm(T) = \int_{\Omega} m_x(T) f(x) \, dx$$

If You subsequently learn that 'x is θ' then in the label semantics framework it would seem most natural for You to update this prior mass assignment on the basis of the constraint $\mathcal{D}_x \in \lambda(\theta)$ in the normal Bayesian manner. This yields the following definition of conditional mass assignment:

DEFINITION 94 *Conditional Mass Assignment on Labels*
$\forall \theta \in LE$ *and* $\forall T \subseteq LA$

$$m_\theta(T) = \begin{cases} \frac{pm(T)}{\sum_{T \in \lambda(\theta)} pm(T)} & : \; T \in \lambda(\theta) \\ 0 & : \; T \notin \lambda(\theta) \end{cases}$$

Hence, one mechanism that You could use to evaluate Your belief that θ is an appropriate description for x given that You know that φ is appropriate to describe x would be to determine a conditional mass assignment m_φ according to definition 94 and then evaluate Your belief that $\mathcal{D}_x \in \lambda(\theta)$ on the basis of this mass assignment. The following theorem shows that such a procedure will lead to the conditional probability $P_{ls}(\theta|\varphi)$ as given in definition 92.

THEOREM 95

$$\forall \theta, \varphi \in LE \; P_{ls}(\theta|\varphi) = \sum_{T \in \lambda(\theta)} m_\varphi(T)$$

Proof
The proof is given for the case where Ω is finite. The proof for the continuous case follows along almost identical lines

$$P_{ls}(\theta|\varphi) = \frac{\sum_{x \in \Omega} \mu_{\theta \wedge \varphi}(x) P(x)}{\sum_{x \in \Omega} \mu_\varphi(x) P(x)} = \frac{\sum_{x \in \Omega} \left(\sum_{T \in \lambda(\theta \wedge \varphi)} m_x(T) \right) P(x)}{\sum_{x \in \Omega} \left(\sum_{T \in \lambda(\varphi)} m_x(T) \right) P(x)}$$

$$= \frac{\sum_{T \in \lambda(\theta \wedge \varphi)} \sum_{x \in \Omega} m_x(T) P(x)}{\sum_{T \in \lambda(\varphi)} \sum_{x \in \Omega} m_x(T) P(x)} = \frac{\sum_{T \in \lambda(\theta \wedge \varphi)} pm(T)}{\sum_{T \in \lambda(\varphi)} pm(T)} \; \text{by definition 93}$$

$$= \sum_{T \in \lambda(\theta) \cap \lambda(\varphi)} \frac{pm(T)}{\sum_{T \in \lambda(\varphi)} pm(T)} = \sum_{T \in \lambda(\theta)} m_\varphi(T) \; \text{by definition 94} \; \square$$

We now investigate the behaviour of the measure P_{ls} with respect to ZIP, CPR1 and CPR2 and show that, in fact, it satisfies all three properties. We begin by considering ZIP. Notice that if $\theta, \varphi \in MLE^{(2)}$ such that $\theta \in LE_1$ and $\varphi \in LE_2$ then automatically we have by theorem 72 (chapter 4) that $\mu_\theta^{(2)}(x_1, x_2) = \mu_\theta(x_1)$ and $\mu_\varphi^{(2)}(x_1, x_2) = \mu_\varphi(x_2)$ as required by ZIP. Now given the conditional independence assumption that $m_{\langle x_1, x_2 \rangle} = m_{x_1} \times m_{x_2}$ (see chapter 4) it follows that:

$$\mu_{\theta \wedge \varphi}^{(2)}(x_1, x_2) = \sum_{\langle T_1, T_2 \rangle \in \lambda^{(2)}(\theta \wedge \varphi)} m_{x_1}(T_1) m_{x_2}(T_2) =$$

$$\sum_{\langle T_1, T_2 \rangle \in \lambda(\theta) \times \lambda(\varphi)} m_{x_1}(T_1) m_{x_2}(T_2)$$

by definition 70 (chapter 4) since $\theta \in LE_1$ and $\varphi \in LE_2$

$$= \left(\sum_{T_1 \in \lambda(\theta)} m_{x_1}(T_1) \right) \times \left(\sum_{T_2 \in \lambda(\varphi)} m_{x_2}(T_2) \right) = \mu_\theta(x_1) \times \mu_\varphi(x_2)$$

Hence,

$$P_{ls}(\theta|\varphi) = \frac{E(\mu_{\theta \wedge \varphi}(x_1, x_2))}{E(\mu_\varphi(x_2))} = \frac{E(\mu_\theta(x_1) \times \mu_\varphi(x_2))}{E(\mu_\varphi(x_2))} =$$

$$\frac{E(\mu_\theta(x_1)) \times E(\mu_\varphi(x_2))}{E(\mu_\varphi(x_2))} = E(\mu_\theta(x_1)) = P(\theta)$$

Now P_{ls} trivially satisfies CPR1 since appropriateness measures satisfy idempotence (i.e. that $\forall x \in \Omega$ and $\forall \theta \in LE$ $\mu_{\theta \wedge \theta}(x) = \mu_\theta(x)$). Also, from theorem 95 we have that P_{ls} satisfies CPR2 as follows:

$$P(\theta|\varphi) + P(\neg\theta|\varphi) = \sum_{T \in \lambda(\theta)} m_\varphi(T) + \sum_{T \in \lambda(\neg\theta)} m_\varphi(T) =$$

$$\sum_{T \in \lambda(\theta) \cup \lambda(\theta)^c} m_\varphi(T) = \sum_{T \subseteq LA} m_\varphi(T) = 1$$

5.6 Conditioning From Mass Assignments in Label Semantics

In the previous sections of this chapter we have investigated what information You can infer regarding the universe Ω from linguistic information describing this domain as provide by an expression in LE. We now consider the case where, instead of taking the form of an expression in LE, Your linguistic knowledge corresponds to a complete mass assignment across the subsets of LA. For example, we might imagine that this mass assignment was obtained as an aggregate description of a database DB as outlined in chapter 4. Given such a mass assignment on label sets what information can You infer regarding the underlying universe Ω? The following definition of a conditional distribution on Ω given a mass assignment on labels was first proposed in [60]:

DEFINITION 96 *Conditional Distribution given a Mass Assignment on Labels Given a mass assignment on labels $m : 2^{LA} \to [0,1]$ then for Ω finite:*

$$\forall x \in \Omega \; P(x|m) = P(x) \sum_{T \subseteq LA} m_x(T) \frac{m(T)}{pm(T)}$$

and for infinite Ω the conditional density is given by:

$$\forall x \in \Omega \; f(x|m) = f(x) \sum_{T \subseteq LA} m_x(T) \frac{m(T)}{pm(T)}$$

where pm is the prior mass assignment as given in definition 93.

This definition can be justified in terms of the theorem of total probability as follows: Suppose that instead of a mass assignment, You knew that the set of appropriate labels for value y selected at random from Ω was certain to be T (i.e. that $\mathcal{D}_y = T$). In this case, You could apply Bayes' theorem to obtain a conditional distribution on Ω as follows:

$$\forall x \in \Omega \, P(y = x | \mathcal{D}_y = T) = \frac{P(y = x) \, P(\mathcal{D}_y = T | y = x)}{P(\mathcal{D}_y = T)} = \frac{P(x) \, m_x(T)}{pm(T)}$$

However, You do not know the exact value of $\mathcal{D}_y$ for a randomly selected value y but rather You only know a distribution for the values of $\mathcal{D}_y$ corresponding to mass assignment m. Hence, in this case You can apply the theorem of total probability to obtain a conditional distribution according to:

$$\forall x \in \Omega \, P(x | m) = \sum_{T \subseteq LA} P(y = x | \mathcal{D}_y = T) \, m(T) =$$

$$\sum_{T \subseteq LA} \frac{P(x) \, m_x(T)}{pm(T)} m(T) = P(x) \sum_{T \subseteq LA} m_x(T) \frac{m(T)}{pm(T)}$$

This definition of $P(\bullet | m)$ has a number of desirable properties. For example, notice that if the mass assignment that You are conditioning on is equal to Your prior mass assignment (i.e. $m = pm$) then the conditional distribution given m is equal to the prior distribution (i.e. $P(x | m) = P(x)$) as would be expected. Also, in the case that You learn that 'x is θ' then definitions 94 and 96 provide a new mechanism by which You can obtain a posterior distribution on Ω. Specifically, You can first determine a conditional mass assignment given θ according to definition 94 (m_θ) and from this evaluate a conditional distribution given this mass assignment according to definition 96 ($P(\bullet | m_\theta)$). For consistency, You would hope that the resulting distribution on Ω would be the same as if You had directly obtain a conditional distribution given θ according to definition 92 (i.e. $P(\bullet | m_\theta) = P(\bullet | \theta)$). The following theorem [67] shows that this is indeed the case:

THEOREM 97

$$\forall \theta \in LA, \, \forall x \in \Omega \, P(x | m_\theta) = P(x | \theta)$$

Proof
We prove the result for finite Ω. The proof for infinite Ω follows along almost

identical lines.

$$\forall \theta \in LE \ \forall x \in \Omega \ P(x|m_\theta) = P(x) \sum_{T \subseteq LA} m_x(T) \frac{m_\theta(T)}{pm(T)} =$$

$$P(x) \sum_{T \in \lambda(\theta)} m_x(T) \frac{m_\theta(T)}{pm(T)}$$

from definition 94

$$= P(x) \sum_{T \in \lambda(\theta)} \frac{m_x(T)}{\sum_{T \in \lambda(\theta)} pm(T)} = P(x) \frac{\sum_{T \in \lambda(\theta)} m_x(T)}{\sum_{T \in \lambda(\theta)} pm(T)}$$

$$= \frac{P(x)\mu_\theta(x)}{\sum_{T \in \lambda(\theta)} pm(T)}$$

Now by definition 93 we have that the denominator in the above expression can be rewritten as:

$$\sum_{T \in \lambda(\theta)} pm(T) = \sum_{T \in \lambda(\theta)} \sum_{x \in \Omega} m_x(T) P(x) = \sum_{x \in \Omega} P(x) \sum_{T \in \lambda(\theta)} m_x(T) =$$

$$\sum_{x \in \Omega} P(x) \mu_\theta(x)$$

Therefore, making the appropriate substitutions we have that:

$$P(x|m_\theta) = \frac{P(x)\mu_\theta(x)}{\sum_{x \in \Omega} P(x)\mu_\theta(x)} = P(x|\theta) \ \text{from definition 92} \ \square$$

The following example illustrates some important features of the conditional density given a mass assignment.

EXAMPLE 98 *Let* $\Omega = [0, 80]$ *and*
$LA = \{young(y), \ middle \ aged(ma), \ old(o)\}$ *with appropriateness measures defined as in chapter 4 example 75. Now consider updating a uniform prior density f given the mass assignment m_{DB} evaluated from the database in example 75. Recall that:*

$$m_{DB}(\{y\}) = 0.1667, \ m_{DB}(\{y, ma\}) = 0.1778, \ m_{DB}(\{ma\}) = 0.2556,$$
$$m_{DB}(\{ma, o\}) = 0.08333, \ m_{DB}(\{o\}) = 0.1667, \ m_{DB}(\emptyset) = 0.5875$$

Assuming a uniform prior then according to definition 93 the prior mass assignment is determined by:

$$\forall T \subseteq LA \ pm(T) = \frac{1}{80} \int_0^{80} m_x(T) \, dx$$

Hence,

$$pm\left(\{y\}\right) = 0.2969, \ pm\left(\{y, ma\}\right) = 0.0469, pm\left(\{ma\}\right) = 0.2344,$$

$$pm\left(\{ma, o\}\right) = 0.0625, \ pm\left(\{o\}\right) = 0.25, \ pm\left(\emptyset\right) = 0.1094$$

From definition 96 the conditional density given m_{DB} is:

$$\forall x \in \Omega \ f\left(x|m_{DB}\right) = \frac{1}{80}\left(\frac{0.1667}{0.2969}m_x\left(\{y\}\right) + \frac{0.1778}{0.0469}m_x\left(\{y, ma\}\right) + \right.$$

$$\frac{0.2556}{0.2344}m_x\left(\{ma\}\right) + \frac{0.08333}{0.0625}m_x\left(\{ma, o\}\right) + \frac{0.1667}{0.25}m_x\left(\{o\}\right) +$$

$$\left.\frac{0.5875}{0.1094}m_x\left(\emptyset\right)\right)$$

The resulting posterior density is shown in figure 5.8. Notice from figure 5.8

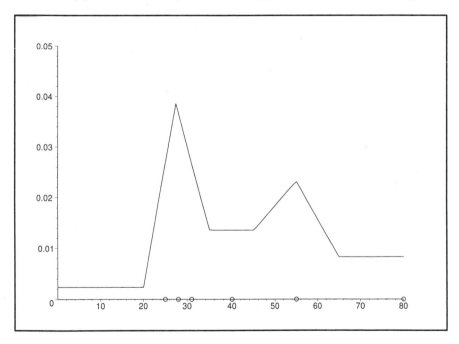

Figure 5.8: Conditional density given m_{DB} (i.e. $f\left(\bullet|m_{DB}\right)$) assuming a uniform prior. The circles along the horizontal axis represent the original values for *age* in DB from which m_{DB} is derived.

that there is a rather exaggerated peak between values 50 and 60. This is due largely to the bi-modal nature of $m_x\left(\emptyset\right)$ as x ranges across Ω. Because of this the occurrence of values between 20 and 30 in DB will also lead to an increase in density between 50 and 60. If the aim is to find a representative density based on DB then clearly this is not desirable. Indeed, in practice, it may be advisable to normalise m_x so that no mass is allocated to the empty *set. For example, one rather natural normalisation method would be for You to reallocate the value of $m_x(\emptyset)$ to the smallest subset of LA to which You give non-zero mass. For example, if*

$$m_x(\{y, ma\}) = m_1, \ m_x(\{ma\}) = m_2, \ m_x(\emptyset) = 1 - m_1 - m_2$$

then the corresponding normalized mass assignment would be:

$$m_x(\{y, ma\}) = m_1, \ m_x(\{ma\}) = 1 - m_1$$

If this process of normalization were carried out for all $x \in \Omega$ then the resulting values for m_x are as shown in figure 5.9 with associated appropriateness measures shown in figure 5.10. Re-evaluating the mass assignment m_{DB} now

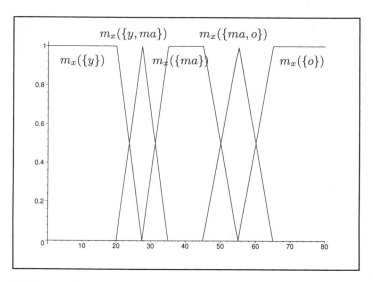

Figure 5.9: Mass assignment values for m_x as x varies from 0 to 80 after normalization

gives:

$$m_{DB}(\{y\}) = 0.0556, \ m_{DB}(\{y, ma\}) = 0.3556, \ m_{DB}(\{ma\}) = 0.2556,$$
$$m_{DB}(\{ma, o\}) = 0.1667, \ m_{DB}(\{o\}) = 0.1667$$

In addition, the prior mass assignment is now given by:

$$pm(\{y\}) = 0.2969, \ pm(\{y, ma\}) = 0.0938, \ pm(\{ma\}) = 0.2344,$$
$$pm(\{ma, o\}) = 0.125, \ pm(\{o\}) = 0.25$$

From this we can derive a new posterior density $f(\bullet|m_{DB})$ as shown in figure 5.11.

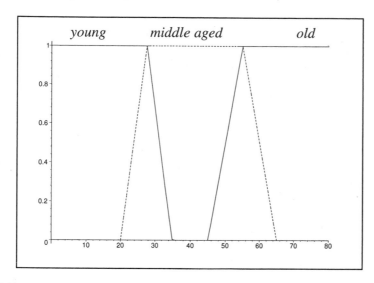

Figure 5.10: Appropriateness measures for labels young, middle aged and old after normalization

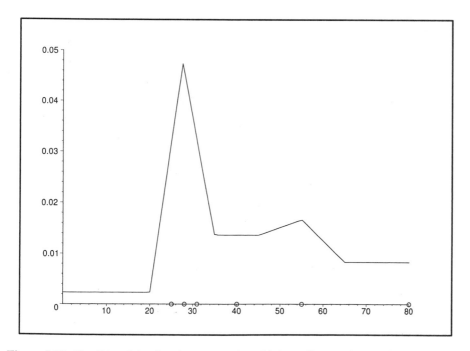

Figure 5.11: Conditional density given m_{DB} (i.e. $f(\bullet|m_{DB})$) assuming a uniform prior and after normalization. The circles along the horizontal axis represent the original values for age in DB from which m_{DB} is derived.

Summary

In this chapter we have overviewed a number of approaches to modelling the information provided by a vague expression 'x is θ' about the underlying universe Ω. In particular, we have discussed Zadeh's work on conditional probability and possibility [111], [115] and have described interpretations of the latter in terms of random sets and imprecise probabilities [104]. As alternatives to Zadeh's work we have proposed definitions for conditional probability and possibility based on mass assignments and on label semantics.

In addition, to the information provided by vague expressions about the underlying universe Ω we also investigated the related question of what information they provide about the likelihood or appropriateness of other expressions. Specifically, we focussed on evaluating possible definitions for the conditional probability $P(\theta|\varphi)$. Four alternative definitions for this conditional probability were considered these being; t-norm based, mass assignment with voter (or α level) independence, mass assignment with voter (or α level) dependence, and label semantics. All definitions were then evaluated in terms of the extent to which they satisfied an independence property ZIP proposed by Zadeh [111], together with two other properties, CPR1 and CPR2. These latter properties effectively require that certainty is preserved for vague concepts, so that $P(\theta|\theta) = 1$ and that probability remains additive for vague concepts, so that $P(\theta|\varphi) + P(\neg\theta|\varphi) = 1$. As evidence for the label semantic approach it is shown that only the label semantic definition of conditional probability satisfies all three of these conditions.

Finally, within the label semantic framework, we considered the case where Your knowledge corresponds to a mass assignment on labels rather than an expression. In this case, we used a argument based on the theorem of total probability to motivate a definition of conditional probability and showed that the latter exhibits a number of desirable and intuitive property. This definition of conditional probability given a mass assignment will be widely applied in chapter 5 on the automated learning of linguistic models from data.

Notes

1 We are abusing notation slightly here since technically α-cuts are defined in terms of membership functions and we are substituting appropriateness measure.

Chapter 6

LEARNING LINGUISTIC MODELS FROM DATA

The title of this volume emphasises that one of our central goals is to develop a formal framework for vague concepts which can be used as a knowledge representation tool in high-level modelling tasks. The rational is that by allowing models to be defined in terms of linguistic expressions we can enhance robustness, accuracy and transparency. In particular, the explicit modelling of uncertainty and imprecision, so often present in complex systems, can improve generalisation and allow for more informative predictions. This uncertainty is not only due to lack of precision or errors in attribute values but is often present in the model itself since the measured attributes may not be sufficient to provide a complete model of the system. Transparency in this context refers to the interpretability of models; that is the ease with which they can be understood by someone who, although well informed regarding the underlying problem, is not an expert on the formal representation framework. Transparent models should allow for a qualitative understanding of the underlying system in addition to giving quantitative predictions of behaviour. In this chapter we shall investigate the effectiveness of label semantics as a modelling framework from the dual perspectives of accuracy and interpretability.

While the development of analytical models may be impractical for many complex systems, there is often data available implicitly describing the behaviour of the system. For example, large companies such as supermarkets, high street stores and banks collect a stream of data relating to the behaviour of their customers. Such data must be analysed to provide models of underlying trends and relationships that can then be used for a range of decision making tasks. From this perspective there is a fundamental need for effective algorithms that can learn linguistic models from data. Here we focus on two types of learning problem, classification and prediction (regression). In the former the goal is to categorise instances into discrete classes while the latter aims to provide an approximation of some real valued function. Two types of linguistic model will be considered: The first will be based on the idea of mass relations as introduced in chapter 4, while the second is a label semantics version of decision trees.

6.1   Defining Labels for Data Modelling

For the models and algorithms described in this chapter, labels provide a means of discretizing a continuous universe (i.e. typically an interval of $\mathbb{R}$). Hence, for a particular data modelling problem it is often desirable to identify appropriateness measures that allow for the most accurate representation of the database. At the same time, since part of the fundamental rational for using linguistic models is to enhance transparency, appropriateness measures should define labels that are reasonably straightforward to interpret. It is arguable that to maximise transparency it would be best to elicit appropriateness measures from experts perhaps by using a voting model as discussed in chapter 3. However, this is unlikely to provide the optimal discretization for data modelling purposes in addition to the fact that such knowledge elicitation is both difficult and time-consuming in practice. In this volume we effectively avoid the complex trade-off between transparency and accuracy, at least at the level of label definitions, by restricting the form of appropriateness measures and then introducing two simple discretization methods.

As a minimal requirement on appropriateness measures we insist that mass is never allocated to the empty set (i.e. $\forall x \in \Omega \, m_x (\emptyset) = 0$). Such a restriction is motivated by the discussion regarding the effect of the empty set on conditional distributions inferred from mass assignments given in example 98 (chapter 5). This assumption is equivalent to the requirement that the labels LA should cover the universe Ω in the sense of the following definition:

DEFINITION 99 *Linguistic Covering*
The labels LA form a linguistic covering of the universe Ω iff

$$\forall x \in \Omega \ \max \left(\{ \mu_L (x) : L \in LA \} \right) = 1$$

For simplicity (and interpretability) in the case that Ω is an interval of $\mathbb{R}$ we also restrict appropriateness measures on labels to those defined by trapezoidal functions as follows:

DEFINITION 100 *Trapezoidal Appropriateness Measures*
A triangular appropriateness measure for label L has the following form: Let

$a, b, c, d \in \mathbb{R}$ *such that* $a \leq b \leq c \leq d$ *then*

$$\mu_L(x) = \begin{cases} 0 & : x \leq a \\ \frac{x-a}{b-a} & : x \in (a, b] \\ 1 & : x \in (b, c] \\ \frac{d-x}{d-c} & : x \in (c, d] \\ 0 & : x > d \end{cases}$$

Discretizing the universe Ω into labels based on trapezoidal appropriateness measures is achieved by using two simple partitioning strategies:

Uniform Appropriateness Measures: According to this approach we first identify the lower and upper bounds, l and u, of the attribute universe and then form a crisp partition by dividing the interval $\Omega = [l, u]$ evenly into a pre-specified number of sub-intervals. A label is then defined for each of these sub-intervals with an appropriateness measure which is equal to 1 over the interval and which tends linearly to zero towards the middle of the adjacent intervals. An illustration of this discretization method is shown in figure 6.1.

Non-Uniform Appropriateness Measures: This discretization method defines labels to fit the distribution of values for the attribute across the database DB. Initially the values from DB are sorted into ascending order, each value now being a potential partition boundary. The interval $\Omega = [l, u]$ is now partitioned into a pre-specified number of sub-intervals such that each interval contains that same number of values from DB. Trapezoidal appropriateness measure are then defined from this crisp partition as in uniform discretization. An illustration of this discretization method is shown in figure 6.2.

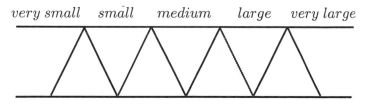

very small   small   medium   large   very large

Figure 6.1: Labels generated using the uniform discretization method

6.2   Bayesian Classification using Mass Relations

Classification is perhaps the most fundamental and widely studied area of machine learning. A classification problem assumes that there is an unknown functional relationship $g : \Omega_1 \times \ldots \times \Omega_k \rightarrow \mathcal{C}$ between a set of model (input) attributes $x_1, \ldots, x_k$ with universes $\Omega_i : i = 1, \ldots, k$ and a discrete classification variable C with universe $\mathcal{C} = \{c_1, \ldots, c_t\}$. Information regarding this

very small   small medium   large   very large

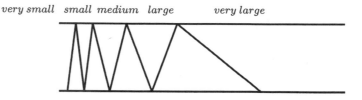

Figure 6.2: Labels generated using the uniform discretization method

function is then provided by a training database containing vectors of input attribute values together with the actual corresponding output class. Let this database be denote by:

$$DB = \{\langle x_1(i), \ldots, x_k(i), C(i) \rangle : i = 1, \ldots, N\}$$

The aim of a classification algorithm is to infer an approximation $\hat{g}$ of the function g from DB. This task is further complicated by the fact that DB may contain noise so that for any data entry, one or more of the values of $x_1, \ldots, x_k$ and C may be erroneous. In the presence of such noise it is necessary to evaluate how well any approximation $\hat{g}$ generalizes from the training DB to other regions of the input-output space not included in this database. This is achieved by measuring the accuracy of the model across a number of test databases different from DB. According to this approach standard statistical techniques can then be used to compare generalization for a number of different classification algorithms.

Bayesian classification algorithms determine $\hat{g}$ on the basis of the conditional probability of C given $x_1, \ldots, x_k$ as inferred from DB using Bayes theorem. More formally:

$$\hat{g}(x_1, \ldots, x_k) = argmax \{P(C|x_1, \ldots, x_k) : C \in \mathcal{C}\} \text{ where}$$

$$P(C|x_1, \ldots, x_k) = \frac{P(x_1, \ldots, x_k|C) P(C)}{\sum_{C \in \mathcal{C}} P(x_1, \ldots, x_k|C) P(C)}$$

Now the estimation of the joint likelihood $P(x_1, \ldots, x_k|C)$ from DB is clearly prone to the curse of dimensionality [8] when k is high. To overcome this, some level of conditional independence is often assumed between $x_1, \ldots, x_k$ given class C. The simplest, and best known, of such algorithms is Naive Bayes (see Lewis [69] for an overview) where complete conditional independence is assumed giving:

$$P(x_1, \ldots, x_k|C) = \prod_{i=1}^{k} P(x_i|C)$$

Using a somewhat weaker version of this assumption Kononenko [55] proposed a form of Semi-Naive Bayes whereby the set of attributes $\{x_1, \ldots, x_k\}$

is partitioned into groups of correlated attributes $S_1, \ldots, S_w$ where $w \leq k$. Conditional independence is then assumed between the attribute groups so that:

$$P(x_1, \ldots, x_k | C) = \prod_{i=1}^{w} P(S_i | C)$$

Clearly then Naive Bayes is a special case of Semi-Naive Bayes where $S_i = \{x_i\} : i = 1, \ldots, k$. We now introduce an algorithm for integrating mass relations (chapter 4) into the Bayesian classifier to obtain a flexible, effective and robust classifier that can provide an intuitive understanding of the function $\hat{g}$. It can also be extended to the case where the model output is a real value rather than a discrete class. In this approach the conditional probabilities $P(x_1, \ldots, x_k | C)$ will be derived from a mass relation that models the relationship between the attributes in $x_1, \ldots, x_k$ for the class C in terms of labels defined in the label semantics framework. Specifically, for each class $c_j \in C$ we can identify a sub-database of DB corresponding to those elements of DB for which $C = c_j$:

$$DB_j = \{\vec{x}(i) \in DB : C(i) = c_j\}$$

We then define a mass relation for this database according to the method for aggregating across a group of objects as described in chapter 4, as follows: Let LA_i be the set of labels describing Ω_i with focal sets $\mathcal{F}_i$ for $i = 1, \ldots, k$, then

$$\forall F_i \in \mathcal{F}_i : i = 1, \ldots, k \ m(F_1, \ldots, F_k | c_j) = \frac{1}{|DB_j|} \sum_{r \in DB_j} \prod_{i=1}^{k} m_{x_i(r)}(F_i)$$

By conditioning on this mass relation according to definition 96 (chapter 5) we can obtain an estimate for $P(x_1, \ldots, x_k | c_j)$ as follows:

$$P(x_1, \ldots, x_k | c_j) = P(x_1, \ldots, x_k | m(\bullet | c_j)) =$$

$$P(x_1, \ldots, x_k) \sum_{F_1 \in \mathcal{F}_1} \cdots \sum_{F_k \in \mathcal{F}_k} \frac{m(F_1, \ldots, F_k | c_j)}{pm(F_1, \ldots, F_k)} \prod_{i=1}^{k} m_{x_i}(F_i) \ \text{and where}$$

$$pm(F_1, \ldots, F_k) = \frac{\prod_{i=1}^{k} \int_{\Omega_i} m_{x_i}(F_i) \, dx_i}{\prod_{i=1}^{k} \int_{\Omega_i} dx_i}$$

based on a uniform prior distribution on Ω_i for $i = 1, \ldots, k$ as given in definition 93 (chapter 5).

EXAMPLE 101 *The Figure of Eight Problem*
We now illustrate the above approach with a simple binary XOR type problem where the object is to classify elements of a subset of $\mathbb{R}^2$ as being either legal, if they lie within a particular rotated eight figure, or illegal if they lie outside it (see

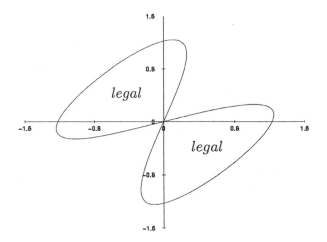

Figure 6.3: Figure of Eight Classification Problem

figure 6.3). More specifically, the objective is to discriminate between the internal and external regions of a figure of eight shape generated according to the parametric equations $x = 2^{-0.5} (\sin 2t - \sin t)$ and $y = 2^{-0.5} (\sin 2t + \sin t)$ where $t \in [0, 2\pi]$. For this problem $\Omega_1 = \Omega_2 = [-1.6, 1.6]$ and we define label sets for x and y as follows:

$$LA_1 = LA_2 =$$
$$\{very\; small(vs),\; small(s),\; medium(m),\; large(l),\; very\; large(vl)\}$$

Identical appropriateness measures were defined for LA_1 and LA_2 corresponding to uniform trapezoidal functions of the type given in example 100. The focal sets for LA_1 and LA_2 then correspond to:

$$\mathcal{F}_1 = \mathcal{F}_2 = \{\{vs\}, \{vs, s\}, \{s\}, \{s, m\}, \{m\}, \{m, l\}, \{l\}, \{l, vl\}, \{vl\}\}$$

The training database DB consisted of 961 examples of x and y co-ordinates together with their associated class, generated from a regular grid across $[-1.6, 1.6]^2$. From this we can infer mass relations conditional on both the legal and illegal classes. Figure 6.4 shows the mass relation given legal in tableau form, also illustrated as a histogram in figure 6.5.

Assuming a uniform prior distribution on $[-1.6, 1.6]^2$ we can condition on the mass relations $m\,(\bullet|legal)$ and $m\,(\bullet|illegal)$ to obtain density functions $f\,(x, y|legal)$ and $f\,(x, y|illegal)$ according to definition 96 (chapter 5). Substituting these densities into Bayes theorem then provides an estimate for $P\,(C|x, y)$. Figure 6.6 shows the conditional density for the legal class derived from the mass relation $m\,(\bullet|legal)$. The classifier was tested on a test database of 2116 unseen example resulting in a classification accuracy of 96.46% on the

	{vs}	{vs,s}	{s}	{s,m}	{m}	{m,l}	{l}	{l,vl}	{vl}
{vl}	0	0	0	0.0023	0.0115	0.0041	0	0	0
{vl,l}	0	0	0.0069	0.0267	0.0369	0.0124	0	0	0
{l}	0	0.0069	0.0304	0.0373	0.0346	0.0069	0	0	0
{l,m}	0.0023	0.0267	0.0373	0.0373	0.0277	0.0028	0.0069	0.0124	0.0041
{m}	0.0115	0.0369	0.0346	0.0277	0.0290	0.0277	0.0346	0.0369	0.0115
{m,s}	0.0041	0.0124	0.0069	0.0028	0.0277	0.0373	0.0373	0.0267	0.0023
{s}	0	0	0	0.0069	0.0346	0.0373	0.0304	0.0069	0
{s,vs}	0	0	0	0.0124	0.0369	0.0267	0.0069	0	0
{vs}	0	0	0	0.0041	0.0115	0.0023	0	0	0

Figure 6.4: Table showing the mass relation $m\left(\bullet|legal\right)$ [87]

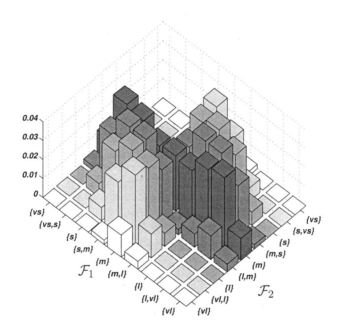

Figure 6.5: Histogram of the mass relation $m\left(\bullet|legal\right)$ [87]

training database DB and 95.94% on the test database (see [87] for more details). Figure 6.7 shows a scatter plot indicating true positives (points correctly classified), false positives (points wrongly classified legal) and false negatives (points wrongly classified illegal).

To alleviate the problems associated with the curse of dimensionality when k is large we can approximate the mass relation $m\left(\bullet|c_j\right)$ using the semi-independence and independence assumption given in definitions 76 and 77 of chapter 4. Partitioning $\{x_1, \ldots, x_k\}$ into attribute groups $S_i : i = 1, \ldots, w$

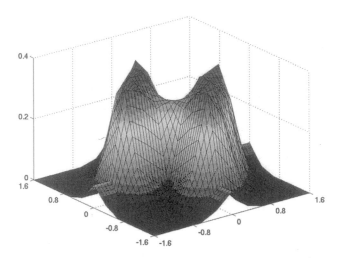

Figure 6.6: Plot showing the density function $f(x, y | m(\bullet | legal))$ derived from the *legal* mass relation [87]

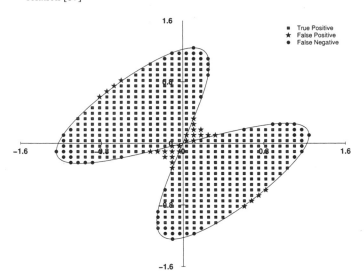

Figure 6.7: Scatter plot showing the classification accuracy of the label semantics Bayesian classifier [87]

then this gives: For $\mathcal{G}_i = \times_{x_j \in S_i} \mathcal{F}_j$ for $i = 1, \ldots, w$ then

$$\forall G_i \in \mathcal{G}_i : i = 1, \ldots, w \; m(G_1, \ldots, G_w | c_j) = \prod_{i=1}^{w} m_i(G_i | c_j)$$

where if w.l.o.g $S_i = \{x_1, \ldots, x_v\}$ and $G_i = \langle F_1, \ldots, F_v \rangle$ then

$$m_i(G_i | c_j) = \frac{1}{|DB_j|} \sum_{r \in DB_j} \prod_{l=1}^{v} m_{x_l(r)}(F_l)$$

In effect then assuming that $m\,(\bullet|c_j)$ is a semi-independent mass relation in this context, is equivalent to a Semi-Naive Bayes model where the conditional probability for attribute group S_i is evaluated from the mass relation $m_i\,(\bullet|c_j)$ according to definition 96 (chapter 5). That is:

$$P\,(S_i|c_j) = P\,(S_i|m_i\,(\bullet|c_j))$$

As noted in chapter 4 one fundamental difficulty with this approach is identifying an effective partition of the attributes. In the following section we investigate methods for learning this partition automatically from DB.

6.2.1   Grouping Algorithms for Learning Dependencies in Mass Relations

Randon [87] and Randon and Lawry [85] and [88] propose a number of search algorithms for identifying the optimal partition of $\{x_1, \ldots, x_k\}$. These are based mainly on the heuristic that attributes should be grouped if and when grouping increases their overall level of importance as an identifier of a given class. The importance of attribute groupings in this context can be quantified using the following measure:

DEFINITION 102 *Importance Measure*
For any input vector S_i the probability of class c_j can be estimated using Bayes theorem where:

$$P(c_j|S_i) = \frac{P(S_i|c_j)|DB_j|}{P(S_i|c_j)|DB_j| + P(S_i|C \neq c_j)|DB - DB_j|}$$

The importance measure for group S_i for class c_j is then defined by:

$$IM(S_i, c_j) = \frac{\sum\limits_{r \in DB_j} P(c_j|S_i(r))}{\sum\limits_{r \in DB} P(c_j|S_i(r))}$$

In the label semantics approach $P(S_i|c_j)$ will be evaluated using the mass relation $m_i\,(\bullet|c_j)$ and $P\,(S_i|C \neq c_j)$ using the mass relation $m_i\,(\bullet|C \neq c_j)$, where the latter is the mass relation describing the attributes in S_i inferred from the sub-database $DB - DB_j$.

The intuition underlying definition 102 is that if the attribute grouping S_i provides an effective means of discriminating the class c_j from other classes then $P\,(c_j|S_i)$ will be relatively low for elements in $DB - DB_j$ and relatively high for elements in DB_j. In this case $IM\,(S_i, c_j)$ will be close to 1.

It is then proposed to guide the search for discriminatory groupings according to the following measure of the improvement in importance obtained by merging two attribute groups. Hence, variable groupings are constructed by starting with single attribute groups $S_i = \{x_i\} : i = 1, \ldots, k$ and then merging groups if and when this leads to an improvement in the ability to discriminate class c_j from other classes.

DEFINITION 103 *(Improvement Measure)*
Suppose we have two subsets of attributes S_1 and S_2 then the improvement in importance obtained by combining them can be calculated as follows:

$$IPM(S_1, S_2, c_j) = 1 - \frac{\min(IM(S_1, c_j), IM(S_2, c_j))}{IM(S_1 \cup S_2, c_j)}$$

The improvement measure can be employed as a heuristic in a range of standard search algorithms whereby the decision to merge attribute grouping S_1 and S_2 is based on whether $IPM(S_1, S_2, c_j)$ exceeds some predefined threshold. The final groupings identified in this manner will be dependent on the nature of the search algorithm and the value of the improvement threshold. Certainly, for all search algorithms it will be necessary to control dimensionality by placing strict limits on the size of attribute groups. Randon [87] proposes depth and breadth first search algorithms guided by the improvement measure as follows:

- **Breadth First Search:** In this algorithm the most important current grouping S_1 is combined with all the other current groupings to identify the grouping S_2 for which the improvement measure $IPM(S_1, S_2, c_j)$ is greatest. Providing this measure exceeds the importance threshold then S_1 and S_2 are merged to give the combined grouping $S_1 \cup S_2$. Next the second most important grouping is tested with the remaining unused groupings and so on. At the next stage the new groupings produced are tested in a similar manner and this continues until either the new grouping gives no more discriminative power than the previous grouping or the maximum grouping size is exceeded. This method provides a fairly extensive search of the space of the partitions, but does limit the structure of the groupings generated.

- **Depth First Search:** In this algorithm the most important grouping S_1 is tested with all other groupings to identify the grouping S_2 for which the improvement measure $IPM(S_1, S_2, c_j)$ is greatest. Providing this measure exceeds the importance threshold then S_1 and S_2 are merged to give the combined grouping $S_1 \cup S_2$. Next $S_1 \cup S_2$ is tested with the unused groupings to see it can be further merged with other groups. This continues until the size of the grouping exceeds the predefined limit or until further merging does not provide significant improvement. The process is repeated with the next most important unused grouping and so on, until all unused groupings have been tested. This allows for a richer structure of groupings but has the disadvantage that some important groupings may be missed.

EXAMPLE 104 *Bristol Image Database*
This example relates to the Bristol image database (see [15] and [71])) con-sisting of 350 colour images of a wide range of urban and rural scenes. The images here are segmented using k-means segmentation (see [15]) on both the training and test images (see [96] p285-290 for full details on the development of the database). From this segmentation a training database of 7,535 and a test database of 3,751 regions were produced. In this version of the database the re-gions are described using 8 attributes with no texture or boundary information. The features and classes described in the database are as follows:

$x_1 = luminance$, $x_2 = red\text{-}green$, $x_3 = yellow\text{-}blue$, $x_4 = size$,
$x_5 = x\text{-}co\text{-}ordinate$ $x_6 = y\text{-}co\text{-}ordinate$, $x_7 = vertical\ orientation$,
$x_8 = horizontal\ orientation$

$c_1 = cloud/mist$, $c_2 = vegetation$, $c_3 = road\ marking$, $c_4 = road\ surface$,
$c_5 = road\ border$, $c_6 = building$, $c_7 = bounding\ objects$, $c_8 = road\ sign$,
$c_9 = signs/poles$, $c_{10} = shadow$ $c_{11} = mobile\ objects$

Randon [87] applied both breadth first and depth first search algorithms to identify attribute groupings on the basis of the improvement measure given in definition 103. Figures 6.8 and 6.9 show the respective search trees for the breadth first and depth first algorithms for the class $c_{10} = shadow$. Clearly

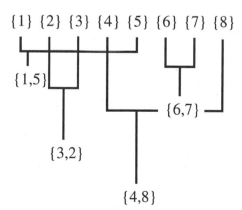

Figure 6.8: The search tree for attribute groupings using a breadth first search guided by the improvement measure (definition 103) [87]

the two search algorithms identify different attribute groupings although it is

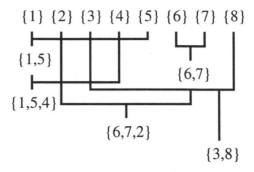

Figure 6.9: The search tree for attribute groupings using a depth first search guided by the improvement measure (definition 103) [87]

interesting to note that both identify clear dependencies between attributes x_1 and x_5 (i.e. luminance and x co-ordinate) and between x_6 and x_7 (i.e. y co-ordinate and vertical orientation).

Randon's [87] model for this problem takes the form of semi-independent mass relations defined over label sets, each containing 4 labels with non-uniform appropriateness measured inferred from DB using the percentile method. The accuracy on the test and training databases for both grouping search algorithms, is shown in figure 6.10 together with results for standard Naive Bayes and the decision tree algorithm C4.5.

	Mass Relations (Breadth)	Mass Relations (Depth)	Naive Bayes	C4.5
Test database	71.37%	71.15%	50.76%	73.13%

Figure 6.10: Results for the Bristol vision database

Randon [87] and Randon and Lawry [88] investigate the performance of the Mass Relation based Bayesian classifier using the semi-independence model described above, on a number of databases taken from the well known UCI machine learning repository [10]. The results obtained were generated from 100 random splits of the database into a training and test sets, with average percentage accuracy and standard deviation shown in figure 6.11 for C4.5, Naive Bayes and the best of the two search algorithms for the mass relation model. Overall we see that the mass relations method performs well on these databases. In fact, comparing the three algorithms using a paired t-test at the 95% level indicates that the mass relation bayesian algorithm, significantly outperforms the other classifiers on the databases listed.

	C4.5	Naive Bayes	Mass Relation
Pima	74.24 ± 2.79	76.58 ± 2.53	77.32 ± 2.60
Sonar	69.95 ± 4.72	67.49 ± 5.08	82.47 ± 3.87
Glass	62.97 ± 5.60	47.33 ± 4.98	67.12 ± 4.41
E.coli	79.71 ± 2.56	83.75 ± 2.30	86.15 ± 1.78

Figure 6.11: Results showing percentage accuracy on UCI databases

6.2.2 Mass Relations based on Clustering Algorithms

As an alternative to the attribute grouping algorithms discussed in the pervious sub-section Lawry [63] and [64] proposes to capture important dependencies between attributes in a mass relational model by using clustering algorithms. Specifically, it is proposed that for each class c_j the corresponding sub-database DB_j should be partitioned into sets of similar elements according to a clustering method (in [64] and [63] the algorithm used is c-means). Suppose that the resulting partition is $\{D_1, \ldots, D_w\}$ where $D_i \cap D_j = \emptyset$ for $i \neq j$ and $D_1 \cup \ldots \cup D_w = DB_j$. Then for each cluster D_i a mass relation $m_{i,j}$ is learnt based on the complete independence assumption given in definition 77 (chapter 4). These are then combined additively to give a conditional mass relation $m(\bullet|c_j)$ as follows:

DEFINITION 105 *Learning Mass Relations using Clustering*

$$\forall F_i \in \mathcal{F}_i : i = 1, \ldots k \ m(F_1, \ldots, F_k|c_j) = \sum_{i=1}^{w} \frac{|D_i|}{|DB_j|} m_{i,j}(F_1, \ldots, F_k)$$

Since the mass relation for each cluster is independent then the curse of dimensionality is avoided, and provided that the elements within a cluster are sufficiently similar so that an independent mass relation provides a good approximation to the joint mass relation then accuracy is maintained. Given such an estimate of $m(\bullet|C)$ for each class C, these are then used to infer conditional distributions for a Bayesian classifier as in the previous sections.

EXAMPLE 106 *Figure of Eight using Clustering*
We now apply the clustering method for learning mass relations to the figure of eight problem described in example 6.3. Notice that for this database an independent mass selection function will not provide an accurate model of either class the since XOR type dependencies between x and y would lead to significant decomposition error. Instead, c-means was used to partition both legal and illegal sub-databases into four clusters. A mass relation was

generated conditional on each class as described above and applying Bayesian classification then gave an accuracy of 95.8% on the test database (as described in example 6.3). This compares with 85.1% if a straightforward independent model is used. Figure 6.12 shows a scatter plot illustrating these results.

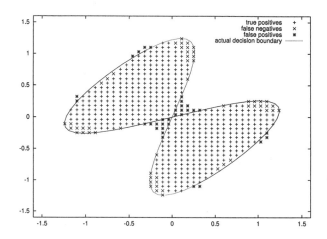

Figure 6.12: Scatter plot showing true positive, false negative and false positive points for the cluster based mass relation classifier on the figure of eight database.

EXAMPLE 107 *Sonar Database*
This database is taken from the UCI repository [10] and was originally used as part of a study into classifying sonar signals using Neural Networks [39]. The objective is to discriminate between sonar signals bounced off a metal cylinder and those bounced off a roughly cylindrical rock. The database consists of 208 patterns, 97 of which are from rocks and 111 from metal cylinders. The training and test sets used for this problem both contain 104 patterns and are identical to those used in [39]. In fact this same database was used for the experiments shown in figure 6.11, however, in this case the data is split in such a way as to take account of aspect-angle dependencies.

Each pattern consists of the values of 60 real valued attributes for a particular object where each attribute represents the energy within a particular frequency band, integrated over a certain period of time. Label sets containing 3 labels were defined for each attribute where the associated appropriateness measure were non-uniform trapezoidal functions. The c-means algorithm, was used to partition both the sub-databases of cylinder and rock patterns with varying values of c for each class. Mass relations conditional on both cylinder and rock classes where inferred as in definition 105.

Figure 6.13 shows the percentage accuracy on the test and training data sets for varying numbers of clusters. Notice that the difference in classification accuracy between the model consisting of 1 cluster per class and that consisting of 4 clusters per class (in this case the optimal) is 11.5% on the training set and 7.7% on the test set. However, the overall accuracy, on the test set in particular, is very sensitive to the number of clusters chosen for both classes. For instance, it is not simply the case that accuracy is a monotonically increasing function of the number of clusters for each class. To see this notice that holding the number of clusters for the cylinder class constant at 1 and increasing the number of clusters for rocks to 4 brings about an overall decrease in classification accuracy of 7.79%. Clearly then the automated learning of optimal cluster numbers for each class remains an open and difficult problem. In the meantime one simplistic solution could be to set an upper bound on prototype numbers and carry out an exhaustive search of the relevant tableau of classification accuracy values on a separate validation set. Interestingly, these results are comparable

	1	2	3	4
1	88.46%	90.38%	91.35%	94.23%
	82.69%	83.65%	81.73%	80.77%
2	81.73%	97.12%	97.12%	98.08%
	74.04%	84.62%	84.62%	83.65%
3	82.69%	97.12%	97.12%	98.08%
	72.12%	82.69%	83.65%	83.65%
4	83.65%	97.12%	98.08%	100%
	75%	90.39%	88.46%	90.39%

Figure 6.13: Table of classification accuracy for training (upper value) and test (lower value) sets for varying numbers of clusters. The number of clusters for cylinders are listed horizontally and the clusters of prototypes for rocks are listed vertically.

with Gorman and Sejnowski [39] who experimented with a back propagation neural network with 60 inputs and up to 24 hidden nodes. In fact their optimal solution was obtained using a network with 12 hidden nodes which gave an accuracy of 90.4% on the test set. These results also compare well with other classifiers such as C4.5, which gives an accuracy of 74.04% on the test set, and Naive Bayes, which gives an accuracy of 73.08%. The accuracy for the mass relational model using attribute grouping was 93.27% on the test set (see [87]).

6.3 Prediction using Mass Relations

Prediction (regression) problems are those where the underlying function g maps into a continuous space rather than a discrete set of classifications. More formally, for prediction we have a set of input attributes $x_1, \ldots, x_k$ with universes $\Omega_i : i = 1, \ldots, k$ and an output attribute x_{k+1} with universe Ω_{k+1} corresponding to some compact interval of $\mathbb{R}$. The aim then is to learn an approximation $\hat{g}$ of the underlying function $g : \Omega_1 \times \ldots \times \Omega_k \to \Omega_{k+1}$ from a database of input vectors $\langle x_1, \ldots x_k \rangle$ together with the associated output $x_{k+1} = g(x_1, \ldots, x_k)$ of the following form:

$$DB = \{ \langle x_1(i), \ldots, x_k(i), x_{k+1}(i) \rangle : i = 1, \ldots, N \}$$

The mass relational learning algorithm for classification problems described in the previous section can be extended to prediction problems by using labels to discretize the output space Ω_{k+1}. This effectively reduces the problem of prediction to one of classification but where the classes correspond to focal sets of the output labels (see figure 6.14). Let $\Omega_i : i = 1, \ldots, k+1$ be discretized using labels $LA_i : i = 1, \ldots, k+1$ with associated focal sets $\mathcal{F}_i : i = 1, \ldots, k+1$. Then for each output focal set $F_{k+1} \in \mathcal{F}_{k+1}$ we can define a conditional mass relation given F_{k+1} as follows:

$$\forall F_i \in \mathcal{F}_i : i = 1, \ldots k \; m(F_1, \ldots, F_k | F_{k+1}) = \frac{\sum_{r \in DB} \prod_{i=1}^{k+1} m_{x_i(r)}(F_i)}{\sum_{r \in DB} m_{x_{k+1}(r)}(F_{k+1})}$$

From conditional mass relations of this kind we can evaluate the conditional

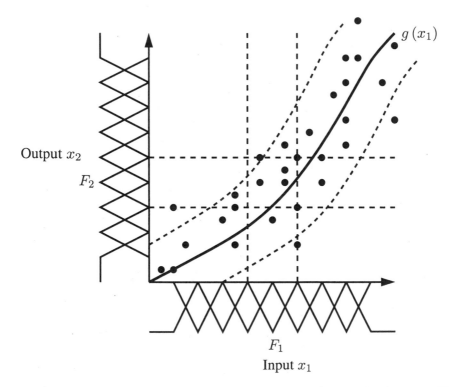

Figure 6.14: Discretization of a prediction problem where $k = 1$, using focal sets. The black dots correspond to data vectors derived from the function $g(x_1)$ but involving some measurement error and other noise.

probability $P(F_{k+1}|x_1, \ldots x_k)$ for any output focal set F_{k+1} given input vector $x_1, \ldots, x_k$ using the Bayesian model, as for classification problems. Notice that by varying F_{k+1} we obtain a mass assignment on the output focal set $\mathcal{F}_{k+1}$ which can then be mapped to a distribution on Ω_{k+1} according to definition 96 (chapter 5) as follows:

$$f(x_{k+1}|x_1, \ldots, x_k) = \sum_{F_{k+1} \in \mathcal{F}_{k+1}} P(F_{k+1}|x_1, \ldots, x_k) f(x_{k+1}|F_{k+1})$$

where assuming a uniform prior distribution on Ω_{k+1}

$$f(x_{k+1}|F_{k+1}) = \frac{m_{x_{k+1}}(F_{k+1})}{\int_{\Omega_{k+1}} m_{x_{k+1}}(F_{k+1}) dx_{k+1}}$$

From this a natural estimate of $x_{k+1} = g(x_1, \ldots, x_k)$ is obtained by taking the expected value of the distribution $f(x_{k+1}|x_1, \ldots, x_k)$ so that:

$$\hat{x}_{k+1} = \hat{g}(x_1, \ldots, x_k) = \int_{\Omega_{k+1}} x_{k+1} f(x_{k+1}|x_1, \ldots, x_k) dx_{k+1} =$$

$$\sum_{F_{k+1} \in \mathcal{F}_{k+1}} P(F_{k+1}|x_1, \ldots, x_k) E(x_{k+1}|F_{k+1}) \quad \text{where}$$

$$E(x_{k+1}|F_{k+1}) = \frac{\int_{\Omega_{k+1}} x_{k+1} m_{x_{k+1}}(F_{k+1}) dx_{k+1}}{\int_{\Omega_{k+1}} m_{x_{k+1}}(F_{k+1}) dx_{k+1}}$$

Hence, the approximation $\hat{g}(x_1, \ldots, x_k)$ is given by the linear combination of expected values of focal set distributions with coefficients corresponding to the probabilities $P(F_{k+1}|x_1, \ldots, x_k)$. In practice, it can often also be effective to use an alternative approximation obtained by simply replacing $E(x_{k+1}|F_{k+1})$ with the mode of the distribution $f(x_{k+1}|F_{k+1})$ corresponding to $argmax\{m_{x_{k+1}}(F_{k+1}) : x_{k+1} \in \Omega_{k+1}\}$.

The semi-independent mass relational model is also straightforward to extend from classification to prediction problems. For output focal set $F_{k+1} \in \mathcal{F}_{k+1}$ the semi-independent mass relation based on groupings $S_1, \ldots, S_w$ is given by:

For $\mathcal{G}_i = \times_{x_j \in S_i} \mathcal{F}_j$ for $i = 1, \ldots, w$ then

$$\forall G_i \in \mathcal{G}_i : i = 1, \ldots, w \; m(G_1, \ldots, G_w | F_{k+1}) = \prod_{i=1}^{w} m_i(G_i | F_{k+1})$$

where if w.l.o.g $S_i = \{x_1, \ldots, x_v\}$ and $G_i = \langle F_1, \ldots, F_v \rangle$ then

$$m_i(G | F_{k+1}) = \frac{1}{|F_{k+1}|} \sum_{r \in DB} \prod_{l=1}^{k+1} m_{x_l(r)}(F_l)$$

and where $|F_{k+1}| = \sum_{r \in DB} m_{x_{k+1}(r)}(F_{k+1})$

The grouping search algorithms for classification can also be applied to prediction problems with a small change to the importance measure given in definition 102 to take account of the uncertainty associated with focal set F_{k+1}:

DEFINITION 108 *Importance Measure for Prediction*
For any input vector S_i the probability of output focal set $F_{k+1} \in \mathcal{F}_{k+1}$ can be estimated using Bayes theorem where:

$$P(F_{k+1}|S_i) = \frac{P(S_i|F_{k+1})|F_{k+1}|}{P(S_i|F_{k+1})|F_{k+1}| + P(S_i|\mathcal{F}_{k+1} - F_{k+1})|\mathcal{F}_{k+1} - F_{k+1}|}$$

The importance measure for group S_i for focal set F_{k+1} is then defined by:

$$IM(S_i, F_{k+1}) = \frac{\sum\limits_{r \in DB} P(F_{k+1}|S_i(r)) m_{x_{k+1}(r)}(F_{k+1})}{\sum\limits_{r \in DB} P(F_{k+1}|S_i(r))}$$

$P(S_i|F_{k+1})$ is evaluated using the mass relation $m_i(\bullet|F_{k+1})$ and $P(S_i|\mathcal{F}_{k+1} - F_{k+1})$ using the mass relation $m_i(\bullet|\mathcal{F}_{k+1} - F_{k+1})$, where the latter is given by

$$m_i(G|\mathcal{F}_{k+1} - F_{k+1}) =$$

$$\frac{1}{|\mathcal{F}_{k+1} - F_{k+1}|} \sum_{r \in DB} \left(\prod_{l=1}^{k} m_{x_l(r)}(F_l) \right) \times \left(\sum_{F \in \mathcal{F}_{k+1} - F_{k+1}} m_{x_{k+1}(r)}(F) \right)$$

and $|\mathcal{F}_{k+1} - F_{k+1}| = \sum\limits_{r \in DB} \sum\limits_{F \in \mathcal{F}_{k+1} - F_{k+1}} m_{x_{k+1}(r)}(F)$

EXAMPLE 109 *Surface Defined by $x_3 = \sin(x_1 \times x_2)$*
In this example a database of 529 vectors was used to describe the function

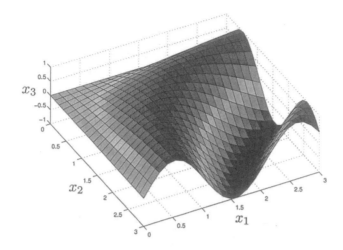

Figure 6.15: The $x_3 = \sin(x_1 \times x_2)$ surface defined by a database of 529 training points.

$x_3 = \sin(x_1 \times x_2)$ *where* $x_1, x_2 \in [0,3]$ *as shown in figure 6.15. Randon and Lawry [86] learnt a mass relational model using 5 labels for* x_1 *and* x_2 *and 7 labels for* x_3. *Non-uniform appropriateness measures were generated as described in the section on classification. The mass relation approximation gave mean a squared error (MSE) of* 0.005 *on a denser test database of* 2209 *examples. Figure 6.16 shows a both the predicted and training surface.*

EXAMPLE 110 *The Sunspot Prediction Problem*
This problem is taken from the Time Series Data Library [48] and contains data on J.R. Wolf's Zürich sunspot relative numbers [49] between the years 1700-1979. In an experiment described in Randon [87] and Randon and Lawry [86] the data was organized as described in [106], apart from the validation set of 35 examples (1921-1955), which was merged into the test set of 24 examples (1956-1979). This is because a validation set is not required for the mass relation learning algorithm as all the required probabilities can be evaluated directly from the training data. Hence, a training set of 209 examples (1712-1920) and a test set of 59 examples (1921-1979) were used. The input attributes were x_{t-12} *to* x_{t-1} *and the output attribute was* x_t *(i.e. one-year-ahead). Hence, the prediction problem was to approximate the function* $x_t = g(x_{t-12}, \ldots, x_{t-1})$. *Each attribute was discretized using 4 labels defined by non-uniform trapezoidal appropriateness measures. The improvement measure threshold was set to 0.9, with a maximum allowed grouping size of 7 attributes. To approximate the underlying function* g, *the mode of the distribution* $f(x_{k+1}|F_{k+1})$ *was used for each focal set* $F_{k+1} \in \mathcal{F}_{k+1}$. *Figure 6.17 shows the results obtained in comparison with those obtained from an* ϵ-*SVR algorithm as implemented by Gunn [40]. The results from the two algorithms are comparable although the*

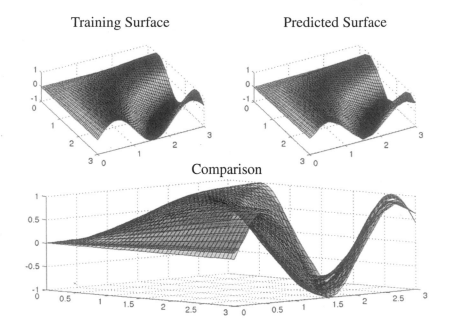

Training Surface   Predicted Surface

Comparison

Figure 6.16: Mass relational model of $x_3 = \sin(x_1 \times x_2)$ [87]

SVR gives marginally better accuracy on the test set with $MSE = 418.126$ as compared to $MSE = 499.659$ for the mass relations. A useful graphical interpretation of prediction results is in terms of scatter plots of the form shown in figure 6.18. In these diagrams each element in the test database is displayed on a two dimensional plot where the x-axis is the actual value and the y-axis is the predicted value. Hence, if a learning algorithm has captured the mapping g completely correctly then all points on such a plot would lie on the line $y = x$. This means that points lying below this line are underestimates of g while those lying above the line are overestimates. For the sunspot problem we can see from figure 6.18 that the SVR tends to systematically under estimate values (figure 6.18 (a)) while the mass relational algorithm shows a more even distribution of errors (figure 6.18 (b)).

6.4   Qualitative Information from Mass Relations

The results given in the previous sections suggest that mass relations can provide effective models with good predictive accuracy. We now analyse these models from the perspective of transparency. In other words, what insight can they give regarding the underlying nature of the system being modelled?

We show that by using the mapping between focal sets and label expressions introduced in chapter 3 (definitions 56 and 59) then mass relations can be used directly to infer qualitative information as expressible within the label

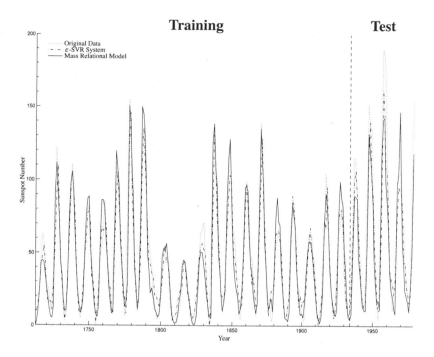

Figure 6.17: Comparison between the ε-SVR prediction, the Mass Relation prediction and the actual sunspot values.

semantics framework. In particular, a set of conditional mass relations on class as described in the previous sections represent a set of quantified linguistic rules describing the function g. For a classification problem given a mass relation $m(F_1, \ldots, F_k | C)$ for each class $C \in \mathcal{C}$ then for each vector of focal sets $\langle F_1, \ldots, F_k \rangle$ where $F_i \in \mathcal{F}_i$ we can derive a weighted rule of the form:

$$\alpha_{F_1} \wedge \alpha_{F_2} \wedge \ldots \wedge \alpha_{F_k} \to C : w$$

where $w = P(C | F_1, \ldots, F_k) = \dfrac{m(F_1, \ldots, F_k | C) |C|}{\sum_{C \in \mathcal{C}} m(F_1, \ldots, F_k | C) |C|}$

and where α_F is the mapping from label sets to expressions given in definition 56 (chapter 3). Similarly, for prediction problems we can replace the class in the consequent of the rule with an output focal set to obtain rules of the form:

$$\alpha_{F_1} \wedge \alpha_{F_2} \wedge \ldots \wedge \alpha_{F_k} \to \alpha_{F_{k+1}} : w$$

where $w = P(F_{k+1} | F_1, \ldots, F_k) = \dfrac{m(F_1, \ldots, F_k | F_{k+1}) |F_{k+1}|}{\sum_{F_{k+1} \in \mathcal{F}_{k+1}} m(F_1, \ldots, F_k | F_{k+1}) |F_{k+1}|}$

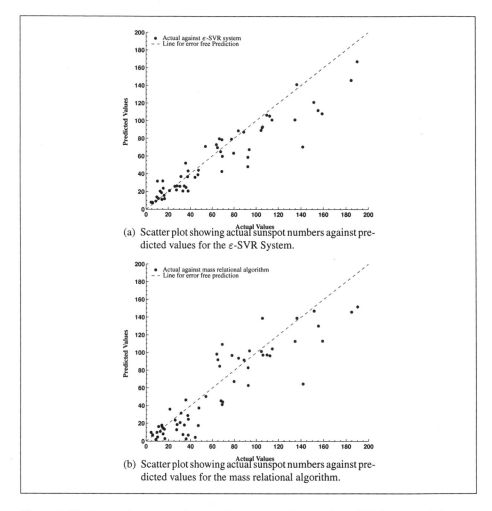

(a) Scatter plot showing actual sunspot numbers against pre-dicted values for the ε-SVR System.

(b) Scatter plot showing actual sunspot numbers against pre-dicted values for the mass relational algorithm.

Figure 6.18: Scatter plots comparing actual sunspot numbers to the ε-SVR System and the mass relational algorithm for the 59 points in the test data set for the years 1921 to 1979.

Such rules can often be simplified by taking into account the relationship between labels as represented by the focal sets $\mathcal{F}$. More specifically, for any focal set $F \in \mathcal{F}$ there is a simplification $\alpha_F^{\mathcal{F}}$ of the atom α_F (definition 56) that only includes the negation of labels that have some overlap with those labels in F. Formally, this simplified mapping from focal sets to atomic label expression is defined as follows:

DEFINITION 111 *Simplified α-mapping*

$$\forall F \in \mathcal{F} \; \alpha_F^{\mathcal{F}} = \left(\bigwedge_{L \in F} L \right) \wedge \left(\bigwedge_{F \in \mathcal{N}(F)} \neg L \right)$$

$$\text{where } \mathcal{N}(F) = \left(\bigcup_{F' \in \mathcal{F} : F' \supseteq F} F' \right) - F$$

It is then straightforward to extend lemma 57 (chapter 3) to show that $\lambda \left(\alpha_F^{\mathcal{F}} \right) \cap \mathcal{F} = \lambda(\alpha_F) = \{F\}$. From definition 111 a natural simplification of the θ mapping (definition 59, chapter 3) from sets of label sets to label expressions can then be defined as follows:

DEFINITION 112 *Simplified θ-mapping*

$$\forall \Psi \subseteq \mathcal{F} \; \theta_\Psi^{\mathcal{F}} = \bigvee_{F \in \Psi} \alpha_F^{\mathcal{F}}$$

It is then a trivial extension of theorem 60 to show that $\lambda \left(\theta_\Psi^{\mathcal{F}} \right) \cap \mathcal{F} = \lambda(\theta_\Psi) = \Psi$

EXAMPLE 113 *Consider the figure of eight classification problem described in example 101. From the table in figure 6.4 we see that*

$$m(\{m, s\}, \{l\} | legal) = 0.0373$$

For this problem it also holds that for the illegal class:

$$m(\{s, m\}, \{l\} | illegal) = 0$$

Hence, we have the rule:

$$\alpha_{\{s,m\}} \wedge \alpha_{\{l\}} \to legal : 1$$

Now for this example we have that

$$LA_1 = LA_2 = \{very\ small(vs), small(s), medium(m), large(l), very\ large(vl)\}$$
and $\mathcal{F}_1 = \mathcal{F}_2 = \{\{vs\}, \{vs, s\}, \{s\}, \{s, m\}, \{m\}, \{m, l\}, \{l\}, \{l, vl\}, \{vl\}\}$

The α-mapping to atomic expressions gives us that:

$$\alpha_{\{s,m\}} = s \wedge m \wedge \neg vs \wedge \neg l \wedge \neg vl \text{ and } \alpha_{\{l\}} = l \wedge \neg vs \wedge \neg s \wedge \neg m \wedge \neg vl$$

However these expressions can be simplified by taking account of the fact that from the focal sets $\mathcal{F}_1$ and $\mathcal{F}_2$ we know that neither small nor medium overlap with very large and that large does not overlap with small or very small. Hence,

$$\alpha_{\{s,m\}}^{\mathcal{F}_1} = s \wedge m \wedge \neg vs \wedge \neg l \text{ and } \alpha_{\{l\}}^{\mathcal{F}_2} = l \wedge \neg m \wedge \neg vl$$

Therefore, we obtain the following linguistic rule:
IF x is *small* and *medium* but not *very small* nor *large* AND y is *large* but not *very large* nor *medium* THEN class is *legal*

Mass relations also provide a mechanism for evaluating the truth of more general label expressions describing the relationship between inputs and outputs in the function g. This can allow for the testing and evaluation of queries and hypotheses regarding g. For example, we might typically wonder how likely it is that the class is C given that the input attributes are described by the multi-dimensional label expression θ. This requires that we evaluate rules of the form:

$$\theta \rightarrow C \text{ where } \theta \in MLE^{(k)}$$

The truth of such rules can be quantified, according to Bayes theorem, by:

$$P(C|\theta) = \frac{P(\theta|C)\,P(C)}{\sum_{C \in \mathcal{C}} P(\theta|C)\,P(C)}$$

were in label semantics:

$$P(\theta|C) = \sum_{\langle F_1,...,F_k \rangle \in \lambda^{(k)}(\theta)} m(F_1, \ldots, F_k|C)$$

Hence, for $C = c_j$ from chapter 4 we have that:

$$P(\theta|C) = \mu_\theta(DB_j) \text{ and therefore } P(C|\theta) = \frac{\mu_\theta(DB_j)\,|DB_j|}{\sum_{C \in \mathcal{C}} \mu_\theta(DB_j)\,|DB_j|}$$

The inverse rule given by:

$$C \rightarrow \theta$$

pertaining to the likelihood that elements of class C can be described as θ is then evaluated directly by:

$$P(\theta|C) = \sum_{\langle F_1,...,F_k \rangle \in \lambda^{(k)}(\theta)} m(F_1, \ldots, F_k|C)$$

For prediction problems it is relevant to investigate linguistic relationships between inputs and output where the latter is also described by label expressions. Typically such a relationship might take the form of a rule such as:

$\theta \rightarrow \varphi$ where $\theta \in MLE^{(k)}$ and $\varphi \in LE_{k+1}$

In the label semantics framework the truth of this rule can be quantified by:

$$P(\varphi|\theta) = \sum_{F_{k+1} \in \lambda(\varphi)} m(F_{k+1}|\theta)$$

where $m(\bullet|\theta)$ is a conditional mass assignment on $\mathcal{F}_{k+1}$ defined by:

$$m(F_{k+1}|\theta) = \frac{P(\theta|F_{k+1})|F_{k+1}|}{\sum_{F_{k+1} \in \mathcal{F}_{k+1}} P(\theta|F_{k+1})|F_{k+1}|} \text{ where}$$

$$P(\theta|F_{k+1}) = \sum_{\langle F_1,\ldots,F_k \rangle \in \lambda^{(k)}(\theta)} m(F_1,\ldots,F_k|F_{k+1})$$

For the inverse rule

$\varphi \rightarrow \theta$ where $\theta \in MLE^{(k)}$ and $\varphi \in LE_{k+1}$

we need to evaluate:

$$P(\theta|\varphi) = \frac{P(\varphi|\theta) P(\theta)}{P(\varphi)}$$

Now $P(\varphi|\theta)$ is as above and

$$P(\theta) = \mu_\theta(DB) = \sum_{\langle F_1,\ldots,F_k \rangle \in \lambda^{(k)}(\theta)} m_{DB}(F_1,\ldots,F_k) \text{ where}$$

$$m_{DB}(F_1,\ldots,F_k) = \frac{1}{|DB|} \sum_{r \in DB} \prod_{i=1}^{k} m_{x_i(r)}(F_i) =$$

$$\sum_{F_{k+1} \in \mathcal{F}_{k+1}} m(F_1,\ldots,F_k|F_{k+1}) \frac{|F_{k+1}|}{|DB|}$$

also

$$P(\varphi) = \mu_\varphi(DB) = \sum_{F_{k+1} \in \lambda(\varphi)} m_{DB}(F_{k+1}) \text{ where}$$

$$m_{DB}(F_{k+1}) = \frac{1}{|DB|} \sum_{r \in DB} m_{x_{k+1}(r)}(F_{k+1}) =$$

$$\sum_{F_1 \in \mathcal{F}_1} \cdots \sum_{F_k \in \mathcal{F}_k} m(F_1,\ldots,F_k|F_{k+1}) \frac{|F_{k+1}|}{|DB|}$$

EXAMPLE 114 *Again consider the simple figure of eight classification problem as described in example 101. Suppose we wish to evaluate the following rule:*

IF x is very large AND y is either very large, or medium but not small THEN *class is illegal*

This translates to the label expression conditional rule:

$$vl_1 \wedge (vl_2 \vee (m_2 \wedge s_2)) \rightarrow illegal$$

Now the multi-dimensional λ-set for the expression $\theta \equiv vl_1 \wedge (vl_2 \vee (m_2 \wedge s_2))$ is given by:

$$\lambda^{(2)}(\theta) \cap (\mathcal{F}_1 \times \mathcal{F}_2) =$$

$$\{\langle \{l, vl\}, \{l, vl\} \rangle, \langle \{l, vl\}, \{vl\} \rangle, \langle \{vl\}, \{l, vl\} \rangle, \langle \{vl\}, \{vl\} \rangle,$$

$$\langle \{l, m\}, \{l, vl\} \rangle, \langle \{l, m\}, \{vl\} \rangle, \langle \{m\}, \{l, vl\} \rangle, \langle \{m\}, \{vl\} \rangle\}$$

$\lambda^{(2)}(\theta)$ *is indicated by the grey cells in the table given in figure 6.19. Figure*

$m(\bullet\|illegal)$	$\{vs\}$	$\{vs, s\}$	$\{s\}$	$\{s, m\}$	$\{m\}$	$\{m, l\}$	$\{l\}$	$\{l, vl\}$	$\{vl\}$
$\{vl\}$	0.0347	0.0208	0.0208	0.0201	0.0170	0.0194	0.0208	0.0208	0.0347
$\{vl, l\}$	0.0208	0.0125	0.0102	0.0035	0.0002	0.0083	0.0125	0.0125	0.0208
$\{l\}$	0.0208	0.0102	0.0023	0	0.0009	0.0102	0.0125	0.0125	0.0208
$\{l, m\}$	0.0201	0.0035	0	0	0.0032	0.0116	0.0102	0.0083	0.0194
$\{m\}$	0.0170	0.0002	0.0009	0.0032	0.0028	0.0032	0.0009	0.0002	0.0170
$\{m, s\}$	0.0194	0.0083	0.0102	0.0116	0.0032	0	0	0.0035	0.0201
$\{s\}$	0.0208	0.0125	0.0125	0.0102	0.0009	0	0.0023	0.0102	0.0208
$\{s, vs\}$	0.0208	0.0125	0.0125	0.0083	0.0002	0.0035	0.0102	0.0125	0.0208
$\{vs\}$	0.0347	0.0208	0.0208	0.0194	0.0170	0.0201	0.0208	0.0208	0.0347

Figure 6.19: This tableau shows the conditional mass relation $m(\bullet|illegal)$ for the figure of eight classification problem. The grey cells indicate the focal set pairs necessary for evaluating the rule $vl_1 \wedge (vl_2 \vee (m_2 \wedge s_2)) \rightarrow illegal$.

6.19 shows the conditional mass relation for the illegal class. From this we can evaluate $P(\theta|illegal)$ as follows:

$$P(\theta|illegal) = m(\{l, vl\}, \{l, vl\}|illegal) + m(\{l, vl\}, \{l\}|illegal) +$$

$$m(\{vl\}, \{l, vl\}|illegal) + m(\{vl\}, \{vl\}|illegal) +$$

$$m(\{m, l\}, \{l, vl\}|illegal) + m(\{m, l\}, \{vl\}|illegal)$$

$$+ m(\{m\}, \{l, vl\}|illegal) + m(\{m\}, \{vl\}|illegal)$$

$$= 0.0208 + 0.0347 + 0.0125 + 0.0208 + 0.0083 + 0.0194 + 0.0002 + 0.0170$$

$$= 0.1337$$

This immediately gives us an evaluation for the inverse rule:

$$illegal \rightarrow vl_1 \wedge (vl_2 \vee (m_2 \wedge s_2))$$

Similarly from the table in figure 6.4 we can evaluate $P(\theta|legal)$ according to:

$P(\theta|legal) = m(\{l, vl\}, \{l, vl\}|legal) + m(\{l, vl\}, \{l\}|legal) +$
$m(\{vl\}, \{l, vl\}|legal) + m(\{vl\}, \{vl\}|legal) + m(\{m, l\}, \{l, vl\}|legal) +$
$m(\{m, l\}, \{vl\}|legal) + m(\{m\}, \{l, vl\}|legal) + m(\{m\}, \{vl\}|legal)$
$= 0 + 0 + 0 + 0 + 0.0124 + 0.0041 + 0.0369 + 0.0115 = 0.0649$

Now the number of legal data elements in DB is given by $|DB_1| = 241$ while the number of illegal elements is $|DB_2| = 720$, hence we can evaluate $P(illegal|\theta)$ as follows:

$$P(illegal|\theta) = \frac{0.1337 \times 720}{0.1337 \times 720 + 0.0649 \times 241} = 0.8602$$

6.5   Learning Linguistic Decision Trees

For classification problems decision trees provide a hierarchical partition of the joint attribute universe $\Omega_1 \times \ldots \times \Omega_k$ into disjoint regions, corresponding to branches, which ideally contain database elements of only one class. In this manner they naturally form conditional rules where the antecedent is a conjunction of constraints on input attributes identifying the class appearing in the consequent. The use of decision trees in machine learning dates back to the early algorithms proposed by Hunt etal. [45] which were then adapted by Quinlan [84] to use an information theoretic heuristic to guide the attribute selection in the generation of the tree.

In this section we propose a form of decision tree with linguistic constraints on attributes expressed within the label semantics framework. As such we aim to extend the decision tree formalism to enable the explicit representation of vagueness and imprecision, as well as to provide a naturally interpretable type of linguistic model. Furthermore, the use of vague constraints on attributes means that it is unlikely that branches will only identify elements of one class. This requires that conditional probabilities for each class are evaluated for every branch which also allows for a more explicit representation of the underlying uncertainty associated with the classification problem. Finally, by discretizing continuous output universes using labels we can extend the application domain of decision trees from classification to prediction problems.

A linguistic decision tree (LDT) is a decision tree where the nodes are attributes from $x_1, \ldots, x_k$ and the edges are label expressions describing each attribute. More formally, supposing that the j'th node at depth d is the attribute x_i then there is a set of label expressions $\mathcal{L}_{j,d} \subseteq LE_i$ forming the edges from

x_i such that:

$$\lambda \left(\bigvee_{\theta \in \mathcal{L}_{j,d}} \theta \right) \supseteq \mathcal{F}_i \text{ and } \forall \theta, \varphi \in \mathcal{L}_{j,d} \; \lambda \left(\theta \wedge \varphi \right) \cap \mathcal{F}_i = \emptyset$$

Also a branch B from a LDT consists of a sequence of expressions $\varphi_1, \ldots, \varphi_m$ where $\varphi_d \in \mathcal{L}_{j,d}$ for some $j \in \mathbb{N}$ for $d = 1, \ldots, m$, augmented by a conditional probability $P(C|B) = P(C|\varphi_1 \wedge \ldots \wedge \varphi_m)$ for every class $C \in \mathcal{C}$. That is:

$$B = \langle \varphi_1, \ldots, \varphi_m, P(c_1|B), \ldots, P(c_t|B) \rangle$$

Hence, every branch B encodes a set of weighted linguistic rules of the form:

$$\varphi_1 \wedge \ldots \wedge \varphi_m \to C : P(C|B) \text{ for each } C \in \mathcal{C}$$

EXAMPLE 115 *Consider a classification problem with attributes* x_1, x_2, x_3 *and classes* $\mathcal{C} = \{c_1, c_2\}$. *Suppose the sets of labels describing the three attributes are:*

$$LA_1 = \{small(s), medium(m), large(l)\}$$
$$LA_2 = \{cold(c), warm(m), hot(h)\}$$
$$LA_3 = \{young(y), middle\ aged(ma), old(o)\}$$

with focal sets:

$$\mathcal{F}_1 = \{\{s\}, \{s, m\}, \{m\}, \{m, l\}, \{l\}\}$$
$$\mathcal{F}_2 = \{\{c\}, \{c, w\}, \{w\}, \{w, h\}, \{h\}\}$$
$$\mathcal{F}_3 = \{\{y\}, \{y, ma\}, \{ma\}, \{ma, o\}, \{o\}\}$$

Consider the LDT shown in figure 6.20 involving all three attributes. In this case:

$$\mathcal{L}_{1,1} = \{small, \neg small\} \subseteq LE_1$$
$$\mathcal{L}_{1,2} = \{cold \wedge \neg warm, warm \wedge \neg hot, hot\} \subseteq LE_2$$
$$\mathcal{L}_{2,2} = \{young, middle\ aged \wedge \neg young \wedge \neg old, old\} \subseteq LE_3$$
$$\mathcal{L}_{3,1} = \{warm, \neg warm\} \subseteq LE_2$$

Now the branch labelled B in figure 6.20 corresponds to

$$\langle \neg s, \neg ma \wedge \neg y \wedge \neg o, w, 0.6, 0.4 \rangle$$

and represents the following 2 rules, both with the same antecedent:

$$\neg small \wedge (middle\ aged \wedge \neg young \wedge \neg old) \wedge warm \to C = c_1 : 0.6$$
$$\neg small \wedge (middle\ aged \wedge \neg young \wedge \neg old) \wedge warm \to C = c_2 : 0.4$$

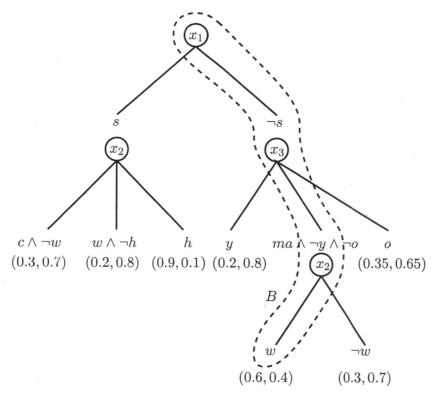

Figure 6.20: Linguistic decision tree involving attributes x_1, x_2, x_3

For classification problems, given a vector of input attribute values $\vec{x} = \langle x_1, \ldots, x_k \rangle$ we need to be able to use LDTs to evaluate the probability $P(C|\vec{x})$ for each class C. The underlying classification mapping $g(\vec{x})$ can then be estimated on the basis of such probabilities, as in the Bayesian model. That is:

$$\hat{g}(\vec{x}) = argmax\{P(C|\vec{x}) : C \in \mathcal{C}\}$$

Consider a LDT consisting of the branches $B_1, \ldots, B_v$ then from the above we know that such branches are exhaustive and exclusive in the following sense. Let $\theta_i \in MLE^{(k)}$ be the conjunction of label expressions appearing in branch B_i. For example, given branch B in figure 6.20 then $\theta = \neg small \wedge (middle\ aged \wedge \neg young \wedge \neg old) \wedge warm$. Now for LDT defined as above it must hold that:

$$\lambda^{(k)}(\theta_1 \vee \ldots \vee \theta_k) \supseteq \mathcal{F}_1 \times \ldots \times \mathcal{F}_k$$

and for $i \neq j$ $\lambda^{(k)}(\theta_i \wedge \theta_j) \cap (\mathcal{F}_1 \times \ldots \times \mathcal{F}_k) = \emptyset$

In this case given an object with attribute values $\vec{x} = \langle x_1, \ldots x_k \rangle$ we can use Jeffrey's rule [50] to determine a probability for each class on the basis of the LDT as follows:

$$P(C|\vec{x}) = \sum_{i=1}^{v} P(C|B_i) \, P(B_i|\vec{x})$$

The conditional probabilities $P(C|B_i)$ are given in the definition of the LDT while in label semantics the conditional probability $P(B_i|\vec{x})$ is interpreted as the probability that θ_i is an appropriate description for $\vec{x}$. Hence,

$$P(B_i|\vec{x}) = \mu_{\theta_i}^{(k)}(\vec{x})$$

Also if $\theta_i = \varphi_1 \wedge \ldots \wedge \varphi_m$ for $m \leq k$ and assuming w.l.o.g that $\varphi_j \in LE_j$ for $j = 1, \ldots, m$ then from the conditional independence assumption described in chapter 4 we have that:

$$\mu_{\theta_i}^{(k)}(\vec{x}) = \prod_{j=1}^{m} \mu_{\varphi_j}(x_j)$$

Suppose for the problem described in example 115 we wish to classify an object with attribute values x_1, x_2, x_3 for which:

$\mu_{small}(x_1) = 0.3,\ \mu_{medium}(x_1) = 1$

$\mu_{warm}(x_2) = 1,\ \mu_{hot}(x_2) = 0.8$

$\mu_{young}(x_3) = 0.6,\ \mu_{middle\ aged}(x_3) = 1$

Assuming a consonant mass selection function then we obtain the following mass assignments on $\mathcal{F}_1, \mathcal{F}_2, \mathcal{F}_3$:

$m_{x_1}(\{s, m\}) = 0.3,\ m_{x_1}(\{m\}) = 0.7$

$m_{x_2}(\{w\}) = 0.2,\ m_{x_2}(\{w, h\}) = 0.8$

$m_{x_3}(\{ma\}) = 0.4,\ m_{x_3}(\{y, ma\}) = 0.6$

For branch B in figure 6.20 $\theta = \neg small \wedge (middle\ aged \wedge \neg young \wedge \neg old) \wedge warm$ and hence:

$$P(\theta|\vec{x}) = \mu_{\theta}^{(3)}(x_1, x_2, x_3) =$$

$\mu_{\neg small}(x_1) \times \mu_{middle\ aged \wedge \neg young \wedge \neg old}(x_3) \times \mu_{warm}(x_2)$

$= 0.7 \times 0.4 \times 0.2 = 0.056$

Similarly, enumerating the branches from left to right in figure 6.20 we have that:

$$P\left(C=c_1|\vec{x}\right) = \sum_{i=1}^{7} P\left(B_i|\vec{x}\right) P(c_1|B_i) =$$

$$0 \times 0.7 + 0.06 \times 0.2 + 0.24 \times 0.9 + 0.42 \times 0.2+$$

$$0.056 \times 0.6 + 0.224 \times 0.3 + 0 \times 0.35 =$$

$$0.4128 P\left(C=c_2|\vec{x}\right) = 1 - P\left(C=c_1|\vec{x}\right) = 0.5872$$

For any LDT there is a corresponding label semantic interpretation where each attribute x_i is replaced by the random set description $\mathcal{D}_{x_i}$ and where each $\theta \in \mathcal{L}_{d,j}$ is replaced with $\lambda\left(\theta\right) \cap \mathcal{F}_i$. Hence, we are replacing linguistic expressions of the form 'x_i is θ' with their label semantics interpretation $\mathcal{D}_{x_i} \in \lambda\left(\theta\right)$ as proposed in chapter 3. For example, figure 6.21 shows the label semantics interpretation of the LDT in figure 6.20. Expressed in this way the most

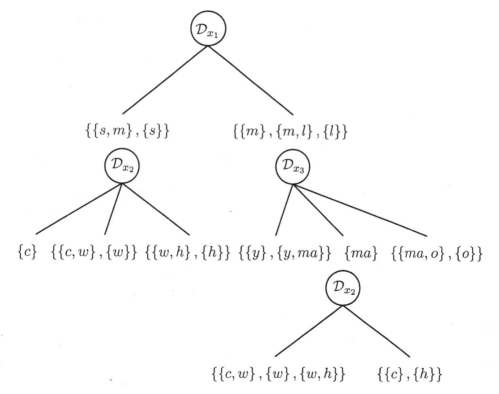

Figure 6.21: Label semantics interpretation of the LDT given in figure 6.20

straightforward LDT is where for node $\mathcal{D}_{x_i}$ each of the edges is a single focal set from $\mathcal{F}_i$. This is the label semantics interpretation of the LDT where each edge is an atom (i.e. where $\mathcal{L}_{j,d} = \left\{ \alpha_F^{\mathcal{F}_i} : F \in \mathcal{F}_i \right\}$). We shall refer to a basic LDT of this kind as a focal element LDT (FELDT). For example, figure 6.22 shows a possible FELDT using the attributes and labels from example 115.

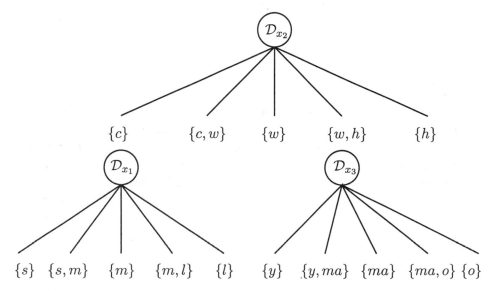

Figure 6.22: A focal element LDT based on the attributes and labels given in example 115

6.5.1   The LID3 Algorithm

Qin and Lawry [82] and [83] introduce the LID3 algorithm, an extension of Quinlan's [84] well-known ID3 algorithm, to learn FELDTs from data. In this approach the class probabilities $P(C|B)$ conditional on a branch B can be evaluated from a database DB in the following manner: In a FELDT branches have the form $B = \langle F_{i_1}, \ldots, F_{i_m} \rangle$ where x_{i_d} is the attribute at the depth d node of B, and $F_{i_d} \in \mathcal{F}_{i_d}$ for $d = 1, \ldots, m$. In this case:

$$P(C = c_j | B) = \frac{\sum_{r \in DB_j} m_{\langle x_{i_1}(r), \ldots, x_{i_m}(r) \rangle} (F_{i_1}, \ldots, F_{i_m})}{\sum_{r \in DB} m_{\langle x_{i_1}(r), \ldots, x_{i_m}(r) \rangle} (F_{i_1}, \ldots, F_{i_m})} =$$

$$\frac{\sum_{r \in DB_j} \prod_{v=1}^{m} m_{x_{i_v}}(r)}{\sum_{r \in DB} \prod_{v=1}^{m} m_{x_{i_v}}(r)}$$

For example, one branch of the FELDT shown in figure 6.22 is $B = \langle \{w, h\}, \{y, ma\} \rangle$. For this branch we have that:

$$P(C = c_j | B) = \frac{\sum_{r \in DB_j} m_{x_2(r)}(\{w, h\}) \times m_{x_3(r)}(\{y, ma\})}{\sum_{r \in DB} m_{x_2(r)}(\{w, h\}) \times m_{x_3(r)}(\{y, ma\})}$$

The numerator in the expression for $P(C|B)$ is proportional to the probability that an element selected at random from DB will have appropriate label sets as given in B while also being of class C. On the other hand the denominator is proportional to the probability that an element of any class picked at random from DB will have appropriate label sets as given in B.

The algorithm for tree generation underlying LID3 is based on essentially the same information theoretic heuristics as ID3 [84] although the evaluation of probability values is adapted to the label semantics context (as, for example, in the evaluation of class probabilities for a branch described above). More formally, the entropy associated with a branch B is defined as follows:

DEFINITION 116 *Branch Entropy*
The entropy of branch B is given by:

$$I(B) = \sum_{C \in \mathcal{C}} P(C|B) \log_2 (P(C|B))$$

Now suppose we wish to extend a branch B by adding a new attribute node x_j, where x_j does not occur in B. This naturally generates a set of branches, one for each of the focal sets in $F_j \in \mathcal{F}_j$. If $B = \langle F_{i_1}, \ldots, F_{i_m} \rangle$ then let $B : F_j$ be the extended branch $\langle F_{i_1}, \ldots, F_{i_m}, F_j \rangle$. Now for each extension $B : F_j$ we can evaluate the branch entropy $I(B : F_j)$ according to definition 116. However, it is unknown which of these extensions $B : F_j$ will describe any particular element described by B, so that we do not know which of these measures is valid. Instead we can evaluate the probability $P(B : F_j|B)$ that an element randomly selected from DB can be described by $B : F_j$ given that it is described by B, as follows:

$$P(B : F_j|B) = \frac{\sum_{r \in DB} \left(\prod_{d=1}^{m} m_{x_{i_d}}(F_{i_d}) \right) \times m_{x_j}(F_j)}{\sum_{r \in DB} \prod_{d=1}^{m} m_{x_{i_d}}(F_{i_d})}$$

From this we can determined the expected branch entropy obtain when the branch B is extended by evaluating attribute x_j according to the following definition:

DEFINITION 117 *Expected Entropy*

$$EI(B : x_j) = \sum_{F_j \in \mathcal{F}_j} I(B : F_j) P(B : F_j|B)$$

From this measure we can quantify the expected information gain (IG) likely to result from extending branch B with attribute x_j as the difference between the branch entropy and the expected entropy so that:

$$IG(B : x_j) = I(B) - EI(B : x_j)$$

The information gain measure underlies both the ID3 and LID3 search strategies. The goal of tree-structured learning models is to make subregions partitioned by branches less 'impure', in terms of the mixture of class labels, than the unpartitioned dataset. For a particular branch, the most suitable free attribute for further expanding, is therefore, the one by which the class 'pureness' is maximally increased as a result of the expansion. This corresponds to selecting the attribute with the maximum information gain. As with ID3, the most informative attribute will form the root of a FELDT, and the tree will expand into branches associated with all possible focal elements of this attribute. For each branch, the free attribute with maximum information gain will then be chosen for the next node, from depth to depth, until the tree reaches a pre-specified maximum depth or the maximum class probability reaches a given threshold probability. The use of termination thresholds in LID3 is another difference from standard ID3. For the discrete attribute decision trees generated by ID3 a branch is only terminated when it is 'pure' (i.e. is only satisfied by elements of one class) or when it already contains all specified attributes as nodes. In FELDTs the inherent uncertainty regarding sets of appropriate labels means that most branches are unlikely to become 'pure' in the sense of a standard decision tree. Instead a branch is terminated if $\max\left(\{P\left(C|B\right):C\in\mathcal{C}\}\right)\geq T$ for some predefined threshold probability $T\in[0,1]$. In addition, to avoid overfitting a limit is set on the depth of branches and LID3 then terminates at that maximal depth.

Qin and Lawry [82] and [83] evaluated the LID3 algorithm by using 14 datasets taken from the UCI repository [10]. These datasets have representative properties of real-world data, such as missing values, multi-classes and unbalanced class distributions. Figure 6.23 gives a summary of the databases used in experiments to evaluate LID3. Unless otherwise stated attributes for each database were discretized by 2 labels with trapezoidal appropriateness measures as in definition 100. For every test the elements of each class were divided evenly between a training and test set, this being repeated randomly 10 times for all the databases. The maximal tree depth was set manually and the results presented below show the best performance of LID3 across a range of depth settings. Also, the threshold probability was set in this case set to $T = 1$ although a more detailed analysis of the overall effected of this learning parameter can be found in [83]. Figure 6.24 gives the results for LID3 using both uniform and non-uniform labels together with those for C4.5, Naive Bayes and Neural Networks all implemented within the WEKA software [109] with default parameter settings. The table in figure 6.24 gives the mean percentage accuracy for each database together with error bounds corresponding to 1 standard deviation from that mean. Figure 6.25 then shows the results of paired t-tests to identify significant differences between the algorithms at the 90% significance level. A 'win' means that the algorithm along the first row of

Index	Database	Size (i.e $|DB|$)	No. Classes (i.e $|\mathcal{C}|$)	Missing Values	No. Attributes (i.e. k)
1	Balance	625	3	No	4
2	Breast-cancer	286	2	Yes	9
3	Breast-w	699	2	No	9
4	Ecoli	336	8	No	8
5	Glass	214	6	No	9
6	Heart-c	303	2	Yes	13
7	Heart-Statlog	270	2	No	13
8	Heptitis	155	2	Yes	19
9	Ionosphere	351	2	No	34
10	Iris	150	3	No	4
11	Liver	345	2	No	6
12	Pima	768	2	No	8
13	Sonar	208	2	No	60
14	Wine	178	3	No	14

Figure 6.23: Summary of UCI databases used to evaluate the LID3 algorithm [83]

the table has a significantly higher accuracy than the corresponding algorithm in the first column; similarly a 'loss' means that the former has a significantly lower accuracy than the latter while a 'tie' means that there is no significant difference between the results of the two algorithms.

Across the databases, all LID3 algorithms outperform C4.5, with LID3-NU achieving the best results with 10 wins, 4 ties and no losses. The performance of the Naive Bayes algorithm and LID3-U is roughly equivalent although LID3-NU outperforms Naive Bayes. Most of the comparisons with Neural Networks result in ties rather than wins or losses. Due to the limited number and types of databases used in these experiments care must be taken in drawing general conclusions regarding the efficacy of LID3 over the other algorithms. However, we can at least conclude that for the experiments in this study LID3 outperforms C4.5 and has equivalent performance to Naive Bayes and Neural Networks. Between different discretization methods LID3 with non-uniform appropriateness measures outperforms LID3 with uniform appropriateness measures.

6.5.2 Forward Merging of Branches

Qin and Lawry [83] also introduce an extension of LID3 to allow for the induction of more general LDTs (rather than only FELDTs). Essentially, a forward merging algorithm groups neighbouring branches into sets of focal elements as the tree is expanded in a breadth first manner. The resulting sets of focal sets can then be converted to label expressions using the simplified θ-mapping given in definition 112. The fact that only neighbouring focal sets are

Index	C4.5	N.B.	N.N.	LID3-U	LID3-NU
1	79.20±1.53	89.46±2.09	90.38±1.18	83.80±1.19	86.23±0.97
2	69.16±4.14	71.26±2.96	66.50±3.48	73.06±3.05	73.06±3.05
3	94.38±1.42	96.28±0.73	94.96±0.80	96.43±0.70	96.11±0.89
4	78.99±2.23	85.36±2.42	82.62±3.18	85.41±1.94	85.59±2.19
5	64.77±5.10	45.99±7.00	64.30±3.38	65.96±2.31	65.87±2.32
6	75.50±3.79	84.24±2.09	79.93±3.99	76.71±3.81	77.96±2.88
7	75.78±3.16	84.00±1.68	78.89±3.05	76.52±3.63	79.04±2.94
8	76.75±4.68	83.25±3.99	81.69±2.48	82.95±2.42	83.08±1.32
9	89.60±2.13	82.97±2.51	87.77±2.88	88.98±2.23	88.01±1.83
10	93.47±3.23	94.53±2.63	95.87±2.70	96.00±1.26	96.40±1.89
11	65.23±3.86	55.41±5.39	66.74±4.89	58.73±1.99	69.25±2.84
12	72.16±2.80	75.05±2.37	74.64±1.41	76.22±1.81	76.54±1.34
13	70.48±0.00	70.19±0.00	81.05±0.00	86.54±0.00	89.42±0.00
14	88.09±4.14	96.29±2.12	96.85±1.57	95.33±1.80	95.89±1.96

Figure 6.24: Accuracy of LID3 based on different discretization methods and three other well-known machine learning algorithms. LID3-U signifies LID3 using uniform discretization and LID3-NU signifies LID3 using non-uniform discretization [83]

	LID3-Uniform vs.	LID3-Non-uniform vs.
C4.5	9 wins-4 ties-1 losses	10 wins-4 ties-0 losses
N.B.	3 wins-8 ties-3 losses	7 wins-4 ties-3 losses
N.N.	5 wins-6 ties-3 losses	5 wins-8 ties-1 losses

Figure 6.25: Summary of t-test comparisons of LID3 based on different discretization methods with three other well-known machine learning algorithms.

considered for merging implicitly assumes some meaningful total ordering on the focal sets describing an attribute. This would certainly seem natural in the case of label sets defined by trapezoidal appropriateness measures as in figures 6.1 and 6.2. In these case if labels $L_1, \ldots, L_n$ have appropriateness measures defined from left to right along the horizontal axis then a natural ordering on the corresponding focal elements would be:

$$\{L_1\} \leq \{L_1, L_2\} \leq \ldots \{L_i\} \leq \{L_i, L_{i+1}\} \leq \{L_{i+1}\} \leq \ldots \leq \{L_n\}$$

Essentially the merging algorithm reviews all branches $B : F$ obtained when branch B is augmented with a new attribute, as the decision tree is generated in a breadth first manner. For example, in figure 6.26 (a) the branch is expanded into five augmented branches with the following focal elements: $\{s\}, \{s, m\}, \{m\}, \{m, l\}, \{l\}$. The branches whose leaves are adjacent focal elements are referred to as *adjacent branches*. If any two adjacent branches have sufficiently similar class probabilities then these two branches give similar classification results and therefore can be merged into one branch in order to obtain a more compact LDT. For example, in figure 6.26 (b) then the two branches $B : F_1$ and $B : F_2$ are merged to form a single augmented branch $B : \{F_1, F_2\}$. The decision criterion for whether or not two adjacent branches

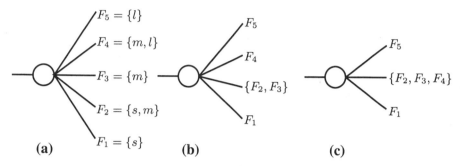

Figure 6.26: An illustration of branch merging in LID3

should be merged is simply the maximal difference between their class probabilities as defined by:

$$\max\left(\{|P\,(C|B_1) - P\,(C|B_2)| : C \in \mathcal{C}\}\right)$$

The branches B_1 and B_2 are then merged in the case that the above measure is less than a merging threshold $T_m \in [0, 1]$.

The incorporation of merged branches into the LDT means that we need to extend the method for calculation conditional class probabilities given a branch. Running LID3 with merging means that, in general, a branch can have the form $B = \langle \Psi_{i_1}, \ldots, \Psi_{i_m} \rangle$ where $\Psi_{i_d} \subseteq \mathcal{F}_{i_d}$ for $d = 1, \ldots m$. Within the label semantics framework the probability of class C given that an element can be described by B can be evaluated from DB according to:

$$P\,(C = c_j|B) = \frac{\sum_{r \in DB_j} \prod_{d=1}^{m} \left(\sum_{F \in \Psi_{i_d}} m_{x_{i_d}}\,(F)\right)}{\sum_{r \in DB} \prod_{d=1}^{m} \left(\sum_{F \in \Psi_{i_d}} m_{x_{i_d}}\,(F)\right)}$$

When merging is applied in LID3 the adjacent branches meeting the merging criteria will be merged and the class probabilities of the resulting branches will be re-evaluated using the above equation. The merging algorithm is then applied recursively to the branches resulting from this first round of merging until no further merging can take place. LID3 then proceeds to the next depth using the information theoretic heuristic, as previously defined. For example, in figure 6.26 (b), the branches with leaves F_2 and F_3 are merged in the first round, while in part (c) the second round of merging results in the combination of branches with leaves $\{F_2, F_3\}$ and F_4.

Qin and Lawry [83] present results from 10 50-50 split experiments using LID3 together with the merging algorithm on the databases described in figure 6.23. In general, we would expect merging to improve the transparency of the LDTs by reducing the number of branches, although this may result in some reduction in accuracy. The merging threshold T_m has a major impact on this transparency accuracy tradeoff. The higher the value of T_m the more likely it is that branch merging will take place as the LDT is being generated, and hence the final number of branches will tend to be smaller. However, if T_m is too high then this may have a significant effect on classification accuracy because merging of incompatible branches may have occurred.

index	n	$T_m = 0$ Acc	$|LDT|$	$T_m = 0.1$ Acc	$|LDT|$	$T_m = 0.2$ Acc	$|LDT|$	$T_m = 0.3$ Acc	$|LDT|$	$T_m = 0.4$ Acc	$|LDT|$
1	2	83.80	77	84.19	51	81.09	25	75.08	10	47.03	1
2	2	73.06	17	71.67	12	71.11	9	59.65	4	61.25	2
3	2	96.43	57	95.80	29	95.74	16	95.63	9	95.49	4
4	3	85.41	345	85.29	445	84.24	203	83.88	104	82.65	57
5	3	65.69	329	62.84	322	64.04	190	44.31	86	35.41	49
6	2	76.71	37	78.68	31	78.55	22	78.42	18	68.49	11
7	3	76.52	31	78.37	35	78.44	23	77.85	12	72.22	7
8	3	82.95	11	81.28	24	80.77	18	80.64	15	80.77	13
9	3	88.98	45	87.90	78	88.47	41	89.20	30	89.20	26
10	3	96.00	21	95.47	23	95.20	18	95.20	14	94.27	10
11	2	58.73	83	56.30	43	55.90	11	57.34	4	57.92	3
12	2	76.12	27	75.31	20	74.45	5	73.85	3	65.10	1
13	2	86.54	615	88.46	516	85.58	337	81.73	93	49.04	6
14	3	95.33	67	93.78	80	94.11	50	93.56	36	89.67	24

Figure 6.27: Comparisons of percentage accuracy Acc and the number of branches (rules) $|LDT|$ with different merging thresholds T_m across a set of UCI datasets. The results for $T_m = 0$ are obtained with $n = 2$ labels and results for other T_m values are obtained with the number of labels n listed in the second column of the table. [83]

However, the results given in figure 6.27 for different values of T_m suggest that it is possible to significantly reduce the tree size (i.e. number of branches) while maintaining good predictive accuracy. For example, figure 6.28 plots the change in percentage accuracy on the test database, and the number of branches (i.e $|LDT|$) against tree depth for different values of T_m on the breast-w data set. Figure 6.28 shows that for this problem accuracy is not greatly influenced by merging, but the number of branches is significantly reduced. For instance, consider the curve for $T_m = 0.3$ (marked +), where merging up to depth 4 only reduces accuracy by 1% as compared to the unmerged case (shown by the $T_m = 0$ curve). In contrast, at this depth, merging reduced the number of branches by roughly 84%.

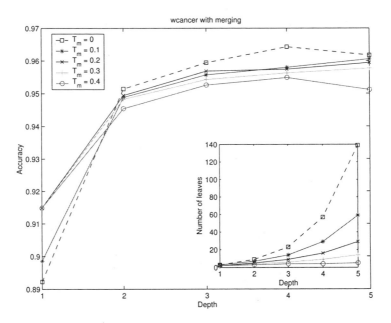

Figure 6.28: The change in accuracy and number of leaves as T_m varies on the breast-w dataset with $n = 2$ labels.

EXAMPLE 118 *Iris Problem*
The Iris Plants database [10] is perhaps the best known database to be found in the pattern recognition literature. It contains 150 instances of Iris plants, described by 4 numeric attributes, quantifying the following physical properties of the flowers: x_1 = sepal length in cm, x_2 = sepal width in cm, x_3 = petal length in cm, and x_4 = petal width in cm. Plants are categorised as being of one of three classes: c_1 = iris setosa, c_2 = iris versicolor and c_3 = iris virginica with the database containing 50 instances of each. To apply LID3

each of the 4 attributes was discretized by 3 labels so that

$$LA_i = \{small\,(s)\,,medium\,(m)\,,large\,(l)\}$$

with corresponding focal sets

$$\mathcal{F}_i = \{\{s\}\,,\{s,m\}\,,\{m\}\,,\{m,l\}\,,\{l\}\}\ for\ i = 1,\dots,3$$

Applying the merging algorithm to a depth of 2 with $T_m = 0.3$ results in the following LDT with an average accuracy 95.2% on the test set as shown in figure 6.29. We can convert this tree to linguistic form by applying the simplified θ-mapping (definition 112) to the sets of focal sets occurring in each branch. This results in the LDT given in figure 6.30.

The LDT in figure 6.30 encodes 24 conditional rules all sharing one of 8 antecedents corresponding to the branches of the tree. A sample of these rules is as follows:

IF petal length is *small but not medium* OR IF petal length is *medium but not large* AND petal width is *small but not medium* THEN class is *iris setosa* : 1

IF petal length is *medium but not large* AND petal width is *large but not medium* THEN class is *iris virginica*: 0.944

IF petal length is *large* AND petal width is *medium but not large* THEN class is *iris versicolor*: 0.842

6.6   Prediction using Decision Trees

LDTs can be extended to prediction problems using a similar approach to mass relations, where the output variable $x_{k+1} = g\,(x_1,\dots,x_k)$ is discretized using labels LA_{k+1} with focal elements $\mathcal{F}_{k+1}$. The form of the LDT is then identical to the classification case, but where the conditional probabilities associated with each branch relate to the output focal sets rather than to classification probabilities. That is, in a prediction LDT a branch has the form:

$$B = \langle \varphi_1,\dots,\varphi_m, P\,(F_{k+1}|B) : F_{k+1} \in \mathcal{F}_{k+1} \rangle$$
where $\varphi_d \in \mathcal{L}_{j,d}$ for some $j \in \mathbb{N}$ for $d = 1,\dots,m$.

The LID3 algorithm can be applied as for classification problems to learn FELDTs but with a minor change to they way in which conditional branch probabilities are calculated, in order to take account of the uncertainty associated with the output focal sets. For instance, if $B = \langle F_{i_1},\dots,F_{i_m} \rangle$ where x_{i_d} is the attribute at the depth d node of B, and $F_{i_d} \in \mathcal{F}_{i_d}$ for $d = 1,\dots,m$ then:

$$P\,(F_{k+1}|B) = \frac{\sum_{r \in DB} \left(\prod_{d=1}^{m} m_{x_{i_d}}(r)\,(F_{i_d}) \right) \times m_{x_{k+1}}(r)\,(F_{k+1})}{\sum_{r \in DB} \prod_{d=1}^{m} m_{x_{i_d}}(r)\,(F_{i_d})}$$

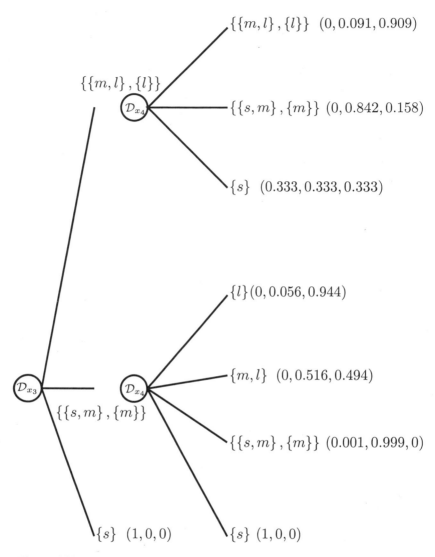

Figure 6.29: LDT tree for the iris databases generated using LID3 with merging

The merging algorithm can also then be implemented along almost identical lines to the classification case but with conditional probabilities adapted as above.

To obtained an estimated output value $\hat{x}_{k+1}$ given inputs $x_1, \ldots, x_k$, a conditional mass assignment on output focal sets can be inferred using Jeffrey's rule as for classification problems so that:

$$P\left(F_{k+1}|x_1, \ldots, x_k\right) = \sum_{B} P\left(F_{k+1}|B\right) P\left(B|x_1, \ldots, x_k\right)$$

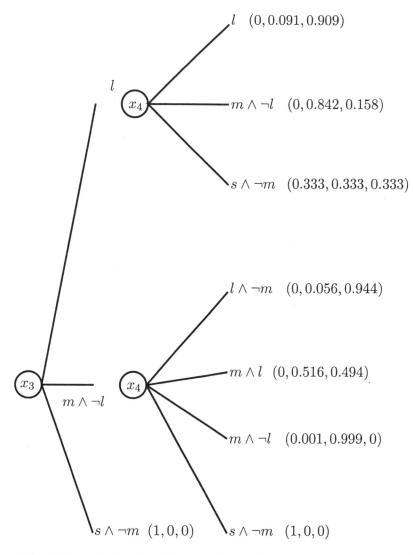

Figure 6.30: LDT tree for the iris databases with sets of focal sets converted to linguistic expressions using the simplified θ-mapping

From this we can derive a conditional distribution $f(x_{k+1}|x_1, \ldots, x_k)$ on Ω_{k+1} and obtain an estimate $\hat{x}_{k+1}$ by taking the expected value as in mass relational prediction models.

EXAMPLE 119 *The Sunspot Problem*
   In this example we apply the LID3 algorithm to the sunspot prediction problem as described in example 110. As before the input attributes were $x_{t-12}, \ldots, x_{t-1}$ predicting x_t, so that $x_t = g(x_{t-12}, \ldots, x_{t-1})$. All input and

the output universes were discretized using 4 non-uniform labels. LID3 was then used to infer an LDT to a maximum depth of 5 with training and test databases as described in example 110. Figure 6.31 shows the results for the LDT, and the ε-SVR [40] together with the actual values on both training and test sets.

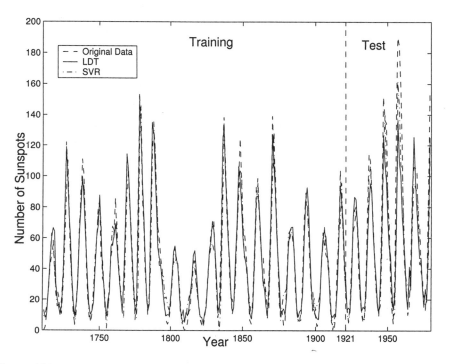

Figure 6.31: Plot showing the sunspot predictions for the LDT and SVR together with the actual values

LID3 was also run on this problem with forward merging. The table in figure 6.32 shows the MSE for LID3 with a range of merging thresholds from $T_m = 0$ (no merging) to $T_m = 0.25$. This clearly illustrates the trade-off between accuracy as measured by MSE and transparency as quantified in this case by tree size (i.e. $|LDT|$). For $T_m = 0.25$ there is a 98.6% reduction in size of the tree but at the cost of a 41.9% increase in MSE. This increase in MSE, however, may over estimate the difference in accuracy between the merged and unmerged models as can be seen from the scatter plot representation of the results as given in figure 6.33, where the difference in predictive performance between the two models appears much less.

Figure 6.32: Prediction results in MSE on the sunspot prediction problem. Results compare the LDT with varying merging thresholds against an SVR and the mass relations method

Prediction Model	MSE		Tree Size
	Training	Test	(LDT only)
Mass relations (independent)	493.91	810.74	-
Mass relations (semi-independent)	134.70	499.66	-
ε-SVR	266.81	418.13	-
LID3	57.824	532.73	5731
LID3 $(T_m = 0.05)$	62.650	549.12	2285
LID3 $(T_m = 0.10)$	69.807	569.60	1493
LID3 $(T_m = 0.15)$	103.82	692.63	757
LID3 $(T_m = 0.20)$	147.20	729.56	204
LID3 $(T_m = 0.25)$	208.29	756.15	81

6.7 Query evaluation and Inference from Linguistic Decision Trees

Clearly the rule-based nature of LDTs means that they provide naturally interpretable models in terms of a set of quantified linguistic rules. However, as with the mass relational approach LDTs can also be used to evaluate general linguistic rules and hypothesis. Specifically, for classification problems consider rules of the form:

$$\theta \to C \text{ where } \theta \in MLE^{(k)} \text{ and } C \in \mathcal{C}$$

Now given a LDT consisting of branches $B_1, \ldots, B_v$ we can use Jeffrey's rule [50] to evaluate the validity of this rule according to:

$$P(C|\theta) = \sum_{i=1}^{v} P(C|B_v) P(B_v|\theta)$$

This requires the evaluation of the conditional probabilities of each branch B_i given the antecedent θ for $i = 1, \ldots, v$ (i.e. $P(B_i|\theta)$). Now suppose that the branch $B = \langle \varphi_1, \ldots, \varphi_m \rangle$ then $P(B|\theta) = P(\varphi_1 \wedge \ldots \wedge \varphi_m|\theta)$ where according to definition 92 (chapter 5):

$$P(\varphi_1 \wedge \ldots \wedge \varphi_m|\theta) = \frac{P(\varphi_1 \wedge \ldots \wedge \varphi_m \wedge \theta)}{P(\theta)}$$

and where the probabilities of label expressions are given by the expected value of their appropriateness measures with respect to an underlying prior probability distribution as in definition 84 (chapter 5). Notice that since we can

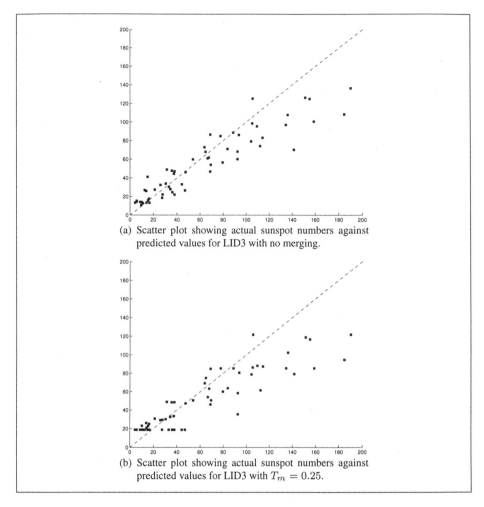

(a) Scatter plot showing actual sunspot numbers against
predicted values for LID3 with no merging.

(b) Scatter plot showing actual sunspot numbers against
predicted values for LID3 with $T_m = 0.25$.

Figure 6.33: Scatter plots comparing actual sunspot numbers to the unmerged LID3 results and
the merged LID3 results for the 59 points in the test data set for the years 1921 to
1979.

evaluate $P(C|\theta)$ for each $C \in \mathcal{C}$ we can classify an object described by θ as
$argmax \{P(C|\theta) : C \in \mathcal{C}\}$.

Provided that, in addition to the LDT, we also know the underlying class
probabilities, $P(C)$ for $C \in \mathcal{C}$, then we can evaluate the inverse rule:

$$C \to \theta$$

by applying Bayes theorem as follows:

$$P(\theta|C) = \frac{P(C|\theta) P(C)}{\sum_{C \in \mathcal{C}} P(C|\theta) P(C)}$$

taking $P(C = c_j) \propto |DB_j|$

In the case of prediction problems rules for evaluation naturally take the form:

$$\theta \to \varphi \text{ where } \theta \in MLE^{(k)} \text{ and } \varphi \in LE_{k+1}$$

Now as in the classification case we can evaluate the conditional probability of each output focal $F_{k+1} \in \mathcal{F}_{k+1}$ set given θ so that:

$$P(F_{k+1}|\theta) = \sum_{i=1}^{v} P(F_{k+1}|B_i) P(B_i|\theta)$$

This then defines a mass assignment on the output focal sets $\mathcal{F}_{k+1}$ and hence we can use label semantics to define the required conditional probability of φ given θ according to:

$$P(\varphi|\theta) = \sum_{F_{k+1} \in \lambda(\varphi)} P(F_{k+1}|\theta)$$

The inverse of this prediction rule is $\varphi \to \theta$ with evaluation $P(\theta|\varphi)$. This is rather problematic to derive from a LDT since the use of Bayes theorem to invert the probability $P(\varphi|\theta)$ requires that we evaluate both $P(\theta)$ and $P(\varphi)$. These probabilities can only be determined from DB and since there is no restriction on θ or φ then in practice this would mean storing the entire database. However, provide the focal element probabilities $P(F_{k+1}) \propto |F_{k+1}|$ are stored for each output focal set $F_{k+1} \in \mathcal{F}_{k+1}$ then rules of the following form can be evaluated as for classification problems by $P(\theta|F_{k+1})$:

$$\varphi \to \theta \text{ where } \theta \in MLE^{(k)}, \varphi \in LE_{k+1} : \lambda(\varphi) \cap \mathcal{F}_{k+1} = \{F_{k+1}\}$$
for some $F_{k+1} \in \mathcal{F}_{k+1}$

EXAMPLE 120 *Consider the problem described in example 115 and associated LDT as shown in figure 6.20. Suppose the prior distribution $f(x_1, x_2, x_3)$ is such that the variables are a prior independent (i.e. $f(x_1, x_2, x_3) = \prod_{i=1}^{3} f_i(x_i)$ -as for the uniform prior) and that based on this distribution the prior mass assignments (definition 93) for each attribute are as follows:*

$pm_1(\{s\}) = 0.1, pm_1(\{s, m\}) = 0.2, pm_1(\{m\}) = 0.2,$
$pm_1(\{m, l\}) = 0.3, pm_1(\{l\}) = 0.2$
$pm_2(\{c\}) = 0.2, pm_2(\{c, w\}) = 0.2, pm_2(\{w\}) = 0.3$
$pm_2(\{w, h\}) = 0.1, pm_2(\{h\}) = 0.2$
$pm_3(\{y\}) = 0.3, pm_3(\{y, ma\}) = 0.1, pm_3(\{ma\}) = 0.3,$
$pm_3(\{ma, o\}) = 0.1, pm_3(\{o\}) = 0.2$

The branches of the LDT in figure 6.20 are, as labelled from left to right:

$B_1 = \langle s, c \wedge \neg w, 0.3, 0.7 \rangle$

$B_2 = \langle s, w \wedge \neg h, 0.2, 0.8 \rangle$

$B_3 = \langle s, h, 0.9, 0.1 \rangle$

$B_4 = \langle \neg s, y, 0.2, 0.8 \rangle$

$B_5 = \langle \neg s, ma \wedge \neg y \wedge \neg o, w, 0.6, 0.4 \rangle$

$B_6 = \langle \neg s, ma \wedge \neg y \wedge \neg o, \neg w, 0.3, 0.7 \rangle$

$B_7 = \langle \neg s, o, 0.35, 0.65 \rangle$

Now suppose we learn that 'x_2 is not cold AND x_3 is middle aged' what are the conditional probabilities of each class? In this case $\theta = \neg c \wedge ma$ and the conditional probability of each branch given θ are evaluated as follows: For B_2 we have that

$$P(B_2|\theta) = P(s \wedge (w \wedge \neg h)\,|\neg c \wedge ma) =$$

$$\frac{P(s \wedge (w \wedge \neg h \wedge \neg c) \wedge ma)}{P(\neg c \wedge ma)}$$

$$= \frac{P(s)\,P(w \wedge \neg h \wedge \neg c)\,P(ma)}{P(\neg c)\,P(ma)} =$$

$$\frac{[pm_1(\{s\}) + pm_1(\{s, m\})]\,[pm_2(\{w\})]}{[pm_2(\{w\}) + pm_2(\{w, h\}) + pm_2(\{h\})]}$$

$$= \frac{0.3 \times 0.3}{0.6} = 0.15$$

Similarly for the remaining branches we have that:

$P(B_1|\theta) = 0$, $P(B_3|\theta) = 0.15$, $P(B_4|\theta) = 0.14$, $P(B_5|\theta) = 0.28$, $P(B_6|\theta) = 0.14$, $P(B_7|\theta) = 0.14$

Applying, Jeffrey's rule we then obtain:

$$P(C = c_1|\theta) = P(B_1|\theta)(0.3) + P(B_2|\theta)(0.2) + P(B_3|\theta)(0.9) +$$

$$P(B_4|\theta)(0.4) + P(B_5|\theta)(0.6) + P(B_6|\theta)(0.3) + P(B_7|\theta)(0.35)$$

$$= 0(0.3) + 0.15(0.2) + 0.15(0.9) +$$

$$0.14(0.2) + 0.28(0.6) + 0.14(0.3) + 0.14(0.35) = 0.4520$$

Similarly,

$$P(C = c_2|\theta) = 1 - P(C = c_1|\theta) = 0.548$$

Hence, an object described as θ would be classified as c_2.

Summary

In this chapter we have proposed two types of linguistic models for both classification and prediction problems. Mass relational methods build conditional models for each class or output focal set using the idea of a mass relation as described in chapter 4. These are then integrated into a Bayesian classification or estimation framework. Linguistic decision trees extend the decision tree formalism to include constraints on attributes defined by label expressions. Classification and prediction is then based on probabilities evaluated using Jeffrey's rule.

The efficacy of these models was investigated from the dual perspectives of predictive accuracy and transparency. Both methods have been shown to give good accuracy across a number of classification and prediction problems with accuracy levels that are comparable with or better than a range of standard machine learning algorithms. In terms of transparency, mass relations generate sets of quantified atomic input-output conditional rules and can be used to evaluate queries or hypotheses expressed as multi-dimensional label expressions. Decision trees have a natural rule representation and the merging algorithm in LID3 allows for the generation of a wider range of descriptive rules than for mass relations. LDTs can also be used for query evaluation although they are slightly more limited than mass relations in this respect because of the difficulty of evaluating rules that are conditional on a constraint on the output attribute.

Chapter 7

FUSING KNOWLEDGE AND DATA

In many knowledge engineering and data modelling applications information is available both in the form of high-level qualitative background knowledge as provided by domain experts and low-level numerical data from experimental studies. To obtain optimal models it is desirable to combine or fuse these two different sources of information. This is particularly true in cases where the data provided is sparse due to the expense or difficulty of experimental trials. The type of fusion method appropriate to a particular problem will be dependent on a number of factors including the nature of the learning algorithms applied to the data and the form of the available background knowledge. In classical analysis the representation of the latter is restricted to one of a few mathematical forms. These include, the choice of parameters or attributes for the learning algorithms, the inclusion of compound parameters as functions of basic parameters, the use of specific parameterised families of probability distributions and the definition of prior distributions in Bayesian methods. However, for many applications the expert knowledge is more naturally represented as 'rules of thumb' expressed as natural language statements. Such statements are likely to be both imprecise and uncertain making their translation into one of the above forms difficult and often inappropriate. Instead, we argue that background knowledge of this kind should be represented in a logical framework, as close to natural language as possible, that also incorporates vagueness, imprecision and uncertainty. Furthermore, the fusion of data with linguistic knowledge is likely to be much easier if both are expressed within the same framework. This latter observation suggests that the data-derived models should also be represented within the same high-level language as the expert knowledge. In this chapter we introduce a fusion method based on label semantics and incorporating the mass relational approach to data modelling described in chapters 4 and 6.

A number of approaches to fusing expert knowledge and data have been proposed in the literature. One family of methods that have received considerable attention are so-called neuro-fuzzy algorithms (See Kruse etal. [56] for an overview). Typically these algorithms involve partitioning each parameter space with fuzzy sets in order to generate a grid of fuzzy cells across the joint parameter space. Every cell identifies a fuzzy conditional rule relating input and output parameters in that region of the space. Available data can then be used to identify which of these cells contain actual measurement values while expert knowledge can identify important or impossible cells. The learning process then corresponds to tuning the fuzzy sets, as characterised by parameterised membership functions, in order to optimise overall performance. Unfortunately this has the disadvantage that model transparency is lost since the optimised fuzzy sets may not be easy to interpret as linguistic labels. Also neuro-fuzzy methods are best suited to classification and prediction tasks where there is a clear measure of model accuracy. In many problems, however, it is necessary to estimate probability distributions over parameter space but where there is no obvious quantifiable measure of the correctness of any such estimate. A typical example of this is reliability analysis where the objective is to estimate the probability of failure of a given engineering system for which very few failures have been observed. The classical approach to fusion in this context is to assume the distribution belongs to a certain parameterised family, as identified by domain experts, and then optimise those parameters to best fit any available data. This can be problematic, however, in that often the available background knowledge is not sufficiently concrete to identify any such family of distributions. Another common approach is Bayesian updating where expert knowledge is encoded as a prior distribution on parameters [52]. Indeed, Bayesian methods have become increasingly popular in the general data modelling context. In particular, Bayesian networks [81] are a powerful modelling tool that allows for causal dependencies, often identified by experts, to be encoded into an estimate of the underlying joint distribution. Traditional Bayesian approaches, however, are limited in that they require background knowledge to be represented either as unique prior distributions or as dependencies between parameters whereas often such knowledge takes the form of imprecise rules of thumb.

7.1   From Label Expressions to Informative Priors

In many modelling problems we may have linguistic knowledge regarding a parameter or group of parameters as provided by an expert or group of experts. In order to combine this with other data-derived knowledge we may wish to transform the linguistic knowledge into an informative prior probability distribution. For the multi-attribute problem suppose that we have linguistic knowledge of the form $\theta \in MLE^{(k)}$ describing attributes $x_1, \ldots, x_k$. This should then be combined with other background information as represented by an initial prior P. In many cases, where θ is the only prior knowledge, then the latter will be the uniform distribution. In this context a natural prior distribution taking into account the prior knowledge θ would correspond to the conditional distribution $P\left(\vec{x}|\theta\right) = P\left(\vec{x}|\left\langle\mathcal{D}_{x_1},\ldots\mathcal{D}_{x_k}\right\rangle \in \lambda^{(k)}\left(\theta\right)\right)$ evaluated as described in chapter 5 from the initial prior. In other words, for the discrete and continuous cases respectively:

$$\forall \vec{x} \in \Omega_1 \times \ldots \Omega_k$$

$$P\left(\vec{x}|\theta\right) = \frac{\mu_\theta^{(k)}\left(\vec{x}\right) P\left(\vec{x}\right)}{\sum_{x_1 \in \Omega_1} \cdots \sum_{x_k \in \Omega_k} \mu_\theta^{(k)}\left(x\right) P\left(\vec{x}\right)}$$

$$f\left(\vec{x}|\theta\right) = \frac{\mu_\theta^{(k)}\left(\vec{x}\right) f\left(\vec{x}\right)}{\int_{\Omega_1} \cdots \int_{\Omega_k} \mu_\theta^{(k)}\left(\vec{x}\right) f\left(\vec{x}\right) d\vec{x}}$$

In a similar manner we can determine an updated mass assignment taking into account the knowledge θ by evaluating m_θ from the prior mass assignment pm according to definition 94 (chapter 5).

On the basis of the above discussion we assume that background knowledge consist of a number of linguistic constraints of the form '$\vec{x}$ is θ_i' for $i = 1, \ldots, v$. This motivates an initial definition of a knowledge base in the label semantics framework, given as follows [68]:

DEFINITION 121 *(Knowledge Base)*
A Knowledge Base is a set of linguistic expressions

$$KB = \{\theta_1, \ldots, \theta_v\} \text{ where } \theta_i \in MLE^{(k)} : i = 1, \ldots, v$$

Given a prior distribution P and a prior mass relation pm (derived from P according to definition 93 chapter 5) we can condition on the knowledge in KB by evaluating the conditional distribution $P\left(\bullet|KB\right) = P\left(\bullet|\bigwedge_{i=1}^{v} \theta_i\right)$ or the conditional mass relation $m_{KB} = m_{\bigwedge_{i=1}^{v} \theta_i}$ as described above. In both these cases we are assuming that KB is consistent so that $\lambda^{(k)}\left(\bigwedge_{i=1}^{v} \theta_i\right) \neq \emptyset$.

EXAMPLE 122 *Consider the problem described in example 74 (chapter 4) where* $LA_1 = \{small_1, medium_1, large_1\}$ *and* $LA_2 = \{small_2, medium_2, large_2\}$ *and where*

$$KB = \{medium_1 \wedge \neg large_1 \rightarrow medium_2, large_1 \rightarrow small_2\}$$

Assuming a uniform initial prior distribution then the conditional density $f\left(x_1, x_2|KB\right)$ *(shown in figure 7.1) is simply a rescaling of the appropriateness measure for* $(m_1 \wedge \neg l_1 \rightarrow m_2) \wedge (l_1 \rightarrow s_2)$ *as shown in figure 4.5 (chapter 4).*

By applying definition 93 (chapter 5) then we obtain the prior mass relation pm *as shown in figure 7.2. Conditioning on KB as described above generates*

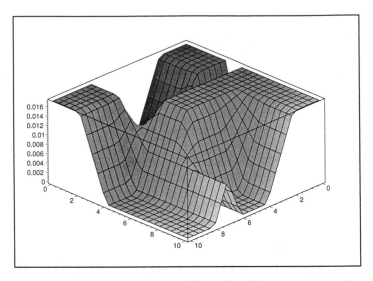

Figure 7.1: Conditional density function $f(x_1, x_2 | KB)$ given knowledge base KB

a conditional mass relation m_{KB} as shown in figure 7.3. Figures 7.2 and 7.3 then provide an interesting perspective on the conditioning process in that the knowledge base KB has the effect of identifying a set of cells of the mass relation tableau as 'possible' and setting all other cells to zero. The mass relation is then renormalised across non-zero cells.

$\times$	$\{s_2\}$	$\{s_2, m_2\}$	$\{m_2\}$	$\{m_2, l_2\}$	$\{l_2\}$	$\emptyset$
$\{s_1\}$	$\frac{1}{16}$	$\frac{1}{80}$	$\frac{3}{40}$	$\frac{1}{80}$	$\frac{1}{16}$	$\frac{1}{40}$
$\{s_1, m_1\}$	$\frac{1}{80}$	$\frac{1}{400}$	$\frac{3}{200}$	$\frac{1}{400}$	$\frac{1}{80}$	$\frac{1}{200}$
$\{m_1\}$	$\frac{3}{40}$	$\frac{3}{200}$	$\frac{9}{100}$	$\frac{3}{200}$	$\frac{3}{40}$	$\frac{3}{100}$
$\{m_1, l_1\}$	$\frac{1}{80}$	$\frac{1}{400}$	$\frac{3}{200}$	$\frac{1}{400}$	$\frac{1}{80}$	$\frac{1}{200}$
$\{l_1\}$	$\frac{1}{16}$	$\frac{1}{80}$	$\frac{3}{40}$	$\frac{1}{80}$	$\frac{1}{16}$	$\frac{1}{40}$
$\emptyset$	$\frac{1}{40}$	$\frac{1}{200}$	$\frac{3}{100}$	$\frac{1}{200}$	$\frac{1}{40}$	$\frac{1}{100}$

Figure 7.2: Prior mass relation pm on $\Omega_1 \times \Omega_2$

In the above we have assumed that knowledge provided by the expert(s) is completely certain. However, in practice this may not be the case and different belief levels may be allocated to each statement. In the current context this corresponds to the association of a subjective probability value with each of the constraints '$\vec{x}$ is θ_i' generating an uncertain knowledge base of the following form [68]:

×	$\{s_2\}$	$\{s_2, m_2\}$	$\{m_2\}$	$\{m_2, l_2\}$	$\{l_2\}$	$\emptyset$
$\{s_1\}$	$\frac{25}{232}$	$\frac{5}{232}$	$\frac{15}{116}$	$\frac{5}{232}$	$\frac{25}{232}$	$\frac{5}{116}$
$\{s_1, m_1\}$	0	$\frac{1}{232}$	$\frac{3}{116}$	$\frac{1}{232}$	0	0
$\{m_1\}$	0	$\frac{3}{116}$	$\frac{9}{58}$	$\frac{3}{116}$	0	0
$\{m_1, l_1\}$	$\frac{5}{232}$	$\frac{1}{232}$	0	0	0	0
$\{l_1\}$	$\frac{25}{232}$	$\frac{5}{232}$	0	0	0	0
$\emptyset$	$\frac{5}{116}$	$\frac{1}{116}$	$\frac{3}{58}$	$\frac{1}{116}$	$\frac{5}{116}$	$\frac{1}{58}$

Figure 7.3: Conditional mass relation m_{KB} on $\Omega_1 \times \Omega_2$

DEFINITION 123 *Uncertain Knowledge Base*

$$KB = \{P(\theta_i) = w_i : i = 1, \ldots, v\}$$

where $\theta_i \in MLE^{(k)}$ and $w_i \in [0, 1]$ for $i = 1, \ldots, v$

DEFINITION 124 *Atoms of a knowledge base*
The atoms of knowledge base KB are all expressions of the form

$$\alpha \equiv \bigwedge_{i=1}^{v} \pm \theta_i \text{ where } +\theta \text{ denotes } \theta \text{ and } -\theta \text{ denotes } \neg\theta.$$

Since there are v constraints in KB there are 2^v atoms. In the sequel these are assumed to be ordered in some arbitrary manner.

Notice that these atoms are mutually exclusive and cover all possible states of the world, hence $\sum_\alpha P(\alpha) = 1$. This means that any uncertain knowledge base KB identifies a unique subset of $[0, 1]^{2^v}$, denoted $V(KB)$, in which each vector corresponds to a probability distribution on the atoms consistent with the constraints in KB.

DEFINITION 125 *Volume of KB*
The volume of KB denoted $V(KB) \subseteq [0, 1]^{2^v}$ such that

$$V(KB) = \left\{ \langle p_1, \ldots, p_{2^v} \rangle \in [0, 1]^{2^v} : \sum_{\alpha_j : \alpha_j \models \theta_i} p_j = w_i, \sum_{j=1}^{2^v} p_j = 1 \right\}$$

The definitions of the fuzzy labels in $LA_i : i = 1, \ldots, k$ and the logical relationship between the expressions $\theta_i : i = 1, \ldots, v$ mean that some atoms are inconsistent and must therefore have zero probability. For example, if the labels *small* and *large* do not overlap then any atoms involving the conjunction *small* $\wedge$ *large* must a priori have probability zero. In view of this, we can naturally define a subset of $[0,1]^{2^v}$ identifying all consistent probability distributions on atoms [68].

DEFINITION 126 *Consistent Volume*

$$V(CON) = \left\{ \langle q_1, \ldots, q_{2^v} \rangle \in [0,1]^{2^v} : \lambda(\alpha_i) = \emptyset \Rightarrow q_i = 0, \sum_{i=1}^{2^v} q_i = 1 \right\}$$

In the sequel we assume w.l.o.g that $\{\alpha_i : \lambda(\alpha_i) = \emptyset\} = \{\alpha_1, \ldots, \alpha_t\}$ *for* $t < 2^v$

From any consistent probability distribution $\vec{q} \in V(CON)$ we can infer a distribution on $\Omega_1 \times \ldots \times \Omega_k$ as follows:

$$P_{\vec{q}}(x_1, \ldots, x_n) = \sum_{i=1}^{2^v} q_i P(x_1, \ldots, x_k | \alpha_i)$$

and similarly a mass relation on $2^{LA_1} \times \ldots \times 2^{LA_k}$ as follows:

$$m_{\vec{q}}(T_1, \ldots, T_n) = \sum_{i=1}^{2^v} q_i m_{\alpha_i}(T_1, \ldots, T_n)$$

In order to estimate the conditional distribution $P(\bullet|KB)$ and conditional mass relation m_{KB} we propose to select an element $\vec{q} \in V(CON)$ that is close, according to some measure, to the elements in $V(KB)$ and then take $P(\bullet|KB) = P_{\vec{q}}(\bullet)$ and $m_{KB} = m_{\vec{q}}$. Here we adopt cross (relative) entropy as our measure of 'closeness' between elements of $V(CON)$ and elements of $V(KB)$ (see Paris [78] for an exposition). Given a distribution $\vec{p} \in V(\mathcal{K})$ this means that we aim to identify the distribution in $\vec{q} \in V(CON)$ that can be obtained by making the minimum change to the information content of $\vec{p}$. The following proposition shows that this 'closest' consistent distribution is obtained by setting the probability of all inconsistent atoms to zero and then renormalising [68].

THEOREM 127 *For any distribution* $\vec{p} \in V(KB)$ *such that* $p_j > 0$ *for* $j = t+1, \ldots, 2^v$, *there exists a unique distribution* $\vec{q} \in V(CON)$ *where* $\vec{q}$ *has minimum cross entropy relative to* $\vec{p}$ *such that:*

$$q_i = \begin{cases} 0 & : \ i = 1, \ldots, t \\ \frac{p_i}{1 - \sum_{j=1}^t p_j} & : \ i = t+1, \ldots, 2^v \end{cases}$$

Proof

$$CE = q_{t+1} \log_2 \left(\frac{q_{t+1}}{p_{t+1}} \right) + \dots$$

$$+ q_{2^v-1} \log_2 \left(\frac{q_{2^v-1}}{p_{2^v-1}} \right) + \left(1 - \sum_{j=t+1}^{2^v-1} q_j \right) \log_2 \left(\frac{(1 - \sum_{j=t+1}^{2^v-1} q_j)}{p_{2^v}} \right)$$

Therefore,

$$\frac{\partial CE}{\partial q_i} = \log_2 \left(\frac{q_i}{p_i} \right) - \log_2 \left(\frac{(1 - \sum_{j=t+1}^{2^v-1} q_j)}{p_{2^v}} \right)$$

Now CE is minimal when $\frac{\partial CE}{\partial q_i} = 0$ and hence when,

$$\frac{p_i}{q_i} = \frac{(1 - \sum_{j=t+1}^{2^v-1} q_j)}{p_{2^v}} \quad : \quad i = t+1, \dots, 2^v - 1$$

Therefore, $\frac{p_i}{q_i} = \frac{p_j}{q_j}$: $j \neq i$ and in particular, $q_i = \frac{q_{t+1}}{p_{t+1}} p_i$: $i = t + 2, \dots, 2^v - 1$

Also

$$\frac{q_{t+1}}{p_{t+1}} = \frac{(1 - q_{t+1} - \sum_{j=t+2}^{2^v-1} q_j)}{p_{2^v}}$$

$$\Rightarrow q_{t+1} = \frac{p_{t+1}(1 - q_{t+1}) - q_{t+1} \sum_{j=v+2}^{2^v-1} p_j}{p_{2^v}} = \frac{p_{t+1} - q_{t+1} \sum_{j=t+1}^{2^v-1} p_j}{p_{2^v}}$$

$$\Rightarrow q_{t+1} p_{2^v} + q_{t+1} \sum_{j=t+1}^{2^v-1} p_j = p_{t+1} \Rightarrow q_{t+1} \left(1 - \sum_{j=1}^{2^v-1} p_j + \sum_{j=t+1}^{2^v-1} p_j \right) = p_{t+1}$$

$$\Rightarrow q_{t+1} \left(1 - \sum_{j=1}^{t} p_j \right) = p_{t+1} \Rightarrow q_{t+1} = \frac{p_{t+1}}{\left(1 - \sum_{j=1}^{t} p_j \right)} \quad \textit{Therefore,}$$

$$q_i = \frac{q_{t+1}}{p_{t+1}} p_i = \frac{p_{t+1}}{\left(1 - \sum_{j=1}^{t} p_j \right)} \frac{p_i}{p_{t+1}} = \frac{p_i}{\left(1 - \sum_{j=1}^{t} p_j \right)} \quad : \quad i = t+1, \dots, 2^v$$

as required. $\square$

If $p_j = 0$ for some $j \in \{t+1, \dots 2^v\}$ but $\sum_{j=t+1}^{2^v} p_j > 0$ then the normalisation identified in theorem 127 can be extended to this case by means of a straightforward limit argument. In the case where $\sum_{j=t+1}^{2^v} p_j = 0$ there is no valid normalisation of this kind. Indeed such a case corresponds to total inconsistency of KB. This occurs, for example, when $KB = \{Pr(\theta_i) = 1 : i = 1, \ldots, v\}$ and $\bigwedge_{i=1}^{v} \theta_i$ is logically inconsistent.

Clearly, theorem 127 identifies a subset of $V(CON)$ of distributions obtained by normalising the distributions in $V(KB)$. However, to select a single representative from this set we need to introduce some more informative inference process. We now consider two methods for selecting a single normalised distribution from $V(CON)$. In the first approach we assume that the expert has allocated the subjective probability values to each expression $\theta_i : i = 1, \ldots, v$ on an independent basis without taking into account the probabilities he is planning to allocate to the other expressions [68].

DEFINITION 128 *(Normalised Independent Solution for KB)*
The normalised independent solution for KB, denoted NIS_{KB}, is given by
$\vec{\hat{p}} \in V(CON)$ *such that:*

$$
\hat{p}_i = \begin{cases}
0 & : \quad i = 1, \ldots, t \\
\dfrac{\prod_{r=1}^{2^v} f_i(w_r)}{1 - \sum_{j=1}^{t} \prod_{r=1}^{2^v} f_i(w_r)} & : \quad i = t+1, \ldots 2^v
\end{cases}
$$

where

$$
f_i(w_j) = \begin{cases}
w_j & : \quad \alpha_i \models \theta_j \\
1 - w_j & : \quad \alpha_i \models \neg\theta_j
\end{cases}
$$

provided $\sum_{j=1}^{t} \prod_{r=1}^{2^v} f_i(w_r) < 1$ *and is undefined otherwise.*

In the second approach we assume that the expert has attempted to be consistent in the allocation of subjective probability values but has failed to properly take into account the underlying interpretation of the label expressions or has failed to detect certain logical inconsistencies. In this case we attempt to identify the distributions in $V(K)$ with which their corresponding normalised distributions in $V(CON)$ have minimum cross entropy [68].

THEOREM 129 *For $\vec{p} \in V(KB)$ such that $\sum_{j=t+1}^{2^v} p_j > 0$, let $\vec{\hat{p}} \in V(CON)$ denote the consistent normalisation of $\vec{p}$ such that*

$$
\hat{p}_i = \begin{cases}
0 & : i = 1, \ldots, t \\
\dfrac{p_i}{1 - \sum_{j=1}^{t} p_j} & : i = t+1, \ldots, 2^v
\end{cases}
$$

Then the cross entropy of $\vec{\hat{p}}$ relative to $\vec{p}$ is minimal across $V(KB)$ if and only if $\sum_{j=1}^{t} p_j$ is minimal.

Proof

$$CE = \sum_{i=t+1}^{2^v} \hat{p}_i \log_2 \left(\frac{\hat{p}_i}{p_i} \right) =$$

$$\sum_{i=t+1}^{2^v} \frac{p_i}{1 - \sum_{j=1}^{t} p_j} log_2 \left(\frac{p_i}{\left(1 - \sum_{j=1}^{t} p_j \right) p_i} \right)$$

$$= \log_2 \left(\frac{1}{1 - \sum_{j=1}^{t} p_j} \right) \sum_{i=t+1}^{2^v} \frac{p_i}{1 - \sum_{j=1}^{t} p_j} =$$

$$\log_2 \left(\frac{1}{1 - \sum_{j=1}^{t} p_j} \right) \frac{\sum_{i=t+1}^{2^v} p_i}{1 - \sum_{j=1}^{t} p_j}$$

$$\log_2 \left(\frac{1}{1 - \sum_{j=1}^{t} p_j} \right) \frac{1 - \sum_{i=1}^{t} p_i}{1 - \sum_{j=1}^{t} p_j} =$$

$$\log_2 \left(\frac{1}{1 - \sum_{j=1}^{t} p_j} \right) = - \log_2 \left(1 - \sum_{j=1}^{t} p_j \right)$$

Now $\log_2(1-x)$ is strictly increasing for $x \in [0,1]$ and hence CE is minimal if and only if $\sum_{j=1}^{t} p_j$ is minimal. $\square$

Hence, from propositions 129 and 127 we see that the set of 'nearest' consistent distributions in $V(CON)$ to those in $V(KB)$, denoted $V(NCKB)$, is given as follows [68]:

DEFINITION 130 *(Nearest Consistent Solutions to KB)*

$$V(NCKB) =$$

$$\left\{ \vec{p} \in V(CON) : \vec{p} \in V(KB) , \ \sum_{j=t+1}^{2^v} p_j > 0 \text{ and } \sum_{j=1}^{t} p_j = \min_{V(KB)} \left(\sum_{j=1}^{t} p_j \right) \right\}$$

As a means of selecting a single representative of $V(NCKB)$ we use maximum entropy to find the solution making the minimum additional assumptions beyond those of closest consistency.

DEFINITION 131 *(Max. Ent. Nearest Consistent Solution to KB)*
The max. ent. nearest consistent solution to KB, denoted $MENCS_{KB}$, is the unique solution in $V(NCKB)$ with maximum entropy.

EXAMPLE 132 *Let x be variable into $\Omega = [0,10]$ with a uniform prior distribution so that $\forall x \in [0,10] \ f(x) = \frac{1}{10}$. For $LA = \{small, medium, large, \}$*

the appropriateness measures are defined by Gaussian functions as shown in figure 7.4. Assuming the consonant msf these generate mass assignments as shown in figure 7.5. In this case the prior mass assignment is given by:

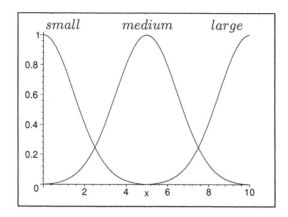

Figure 7.4: Gaussian appropriateness measures for *small*, *medium*, and *large*

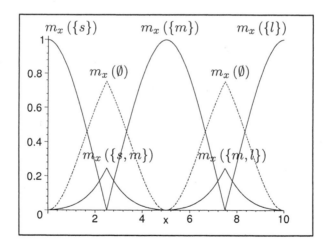

Figure 7.5: Mass assignments on 2^{LA} generated by the appropriateness measures in figure 7.4 under the consonant msf.

$$pm = \{s\} : 0.1528, \{s, m\} : 0.0338, \{m\} : 0.3056, \{m, l\} : 0.0338,$$
$$\{l\} : 0.1528, \emptyset : 0.3212$$

Now suppose we have an uncertain knowledge base:

$$KB = \{P(small \wedge medium) = 0.7, \ P(medium) = 0.5, \ P(\neg large) = 0.4\}$$

According to definition 124 the atoms for this knowledge base are:

$\alpha_1 = (s \wedge m) \wedge m \wedge l,\ \alpha_2 = (s \wedge m) \wedge m \wedge \neg l,\ \alpha_3 = (s \wedge m) \wedge \neg m \wedge l,$
$\alpha_4 = (s \wedge m) \wedge \neg m \wedge \neg l,\ \alpha_5 = \neg(s \wedge m) \wedge m \wedge l,\ \alpha_6 = \neg(s \wedge m) \wedge m \wedge \neg l,$
$\alpha_7 = \neg(s \wedge m) \wedge \neg m \wedge l,\ \alpha_8 = \neg(s \wedge m) \wedge \neg m \wedge \neg l$

So that a probability distribution over the atoms has the following form:

$P((s \wedge m) \wedge m \wedge l) = p_1,\ P((s \wedge m) \wedge m \wedge \neg l) = p_2$
$P((s \wedge m) \wedge \neg m \wedge l) = p_3,\ P((s \wedge m) \wedge \neg m \wedge \neg l) = p_4$
$P(\neg(s \wedge m) \wedge m \wedge l) = p_5,\ P(\neg(s \wedge m) \wedge m \wedge \neg l) = p_6$
$P(\neg(s \wedge m) \wedge \neg m \wedge l) = p_7,\ P(\neg(s \wedge m) \wedge \neg m \wedge \neg l) = p_8$

then

$$V(KB) = \{\langle p_1, \ldots, p_8 \rangle \in [0,1]^8 : p_1 + p_2 + p_3 + p_4 = 0.7,$$

$$p_1 + p_2 + p_5 + p_6 = 0.5, p_1 + p_3 + p_5 + p_7 = 0.6, \sum_{i=1}^{8} p_i = 1\}$$

However, given the definition of the labels small, medium and large (see figures 7.4 and 7.5) we have that:

$$\lambda((s \wedge m) \wedge m \wedge l) \cap \mathcal{F} = \lambda((s \wedge m) \wedge \neg m \wedge l) \cap \mathcal{F} = \lambda((s \wedge m) \wedge \neg m \wedge \neg l) \cap \mathcal{F} = \emptyset$$

In the first case this inconsistency is due to the fact that at no point do small, medium and large all overlap. In the latter two cases the expressions $(s \wedge m) \wedge \neg m \wedge l$ and $(s \wedge m) \wedge \neg m \wedge \neg l$ are logically inconsistent. Hence,

$$V(CON) = \{\langle q_1, \ldots, q_8 \rangle \in [0,1]^8 : q_1 = q_3 = q_4 = 0, \sum_{i=1}^{8} q_i = 1\}$$

The normalised independent solution is given by:

$$\vec{q} = \langle 0, 0.31818, 0, 0, 0.204545, 0.136363, 0.204545, 0.136363 \rangle$$

so that (as shown as the dashed line in figure 7.6) the NIS_{KB} density function is given by:

$f(x|KB) = 0.31818f(x|(s \wedge m) \wedge m \wedge \neg l)+$
$0.204545f(x|\neg(s \wedge m) \wedge m \wedge l)+$
$0.136363f(x|\neg(s \wedge m) \wedge m \wedge \neg l)+$
$0.204545f(x|\neg(s \wedge m) \wedge \neg m \wedge l)+$
$0.136363f(x|\neg(s \wedge m) \wedge \neg m \wedge \neg l)$
$= 0.31818f(x|s \wedge m \wedge \neg l) + 0.204545f(x|\neg s \wedge m \wedge l)+$
$0.136363f(x|\neg s \wedge m \wedge \neg l) + 0.204545f(x|\neg m \wedge l)+$
$0.136363f(x|\neg m \wedge \neg l)$

In order, to determine $V(NCKB)$ we note that this corresponds to the set

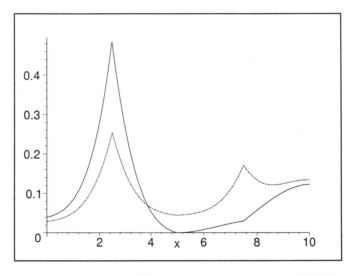

Figure 7.6: Densities generated by the NIS_{KB} (dashed line) and the $MENCS_{KB}$ (solid line) methods, assuming a uniform initial prior.

of consistent normalisations of elements $\vec{p} \in V(KB)$ where $p_1 + p_3 + p_4$ are minimal. Now since $\vec{p} \in V(KB)$ we have that $p_4 = 0.7 - p_1 - p_2 - p_3$ and hence $p_1 + p_3 + p_4 = 0.7 - p_2$. Clearly, this expression is minimal when p_2 is maximal. Again, since $\vec{p} \in V(KB)$ it follows that $p_2 \leq 0.5$. In fact there is a subset of solutions such that $p_2 = 0.5$, $p_1 = p_5 = 0$ and $p_3 \in [0.3, 0.6]$. Hence,

$$V(NCKB) = \left\{ \left\langle 0, 0.625, 0, 0, 0, 0, \frac{0.6 - p_3}{0.8}, \frac{p_3 - 0.3}{0.8} \right\rangle : p_3 \in [0.3, 0.6] \right\}$$

Hence, $V(NCKB)$ generates a family of densities characterised by (see figure 7.7):

$$f(x|KB) = 0.625f(x|s \wedge m \wedge \neg l) + \frac{0.6 - p_3}{0.8}f(x|\neg m \wedge l)$$

$$+\frac{p_3 - 0.3}{0.8}f(x|\neg m \wedge \neg l) : p_3 \in [0.3, 0.6]$$

The maximum entropy solution in $V(NCKB)$ is when $p_3 = 0.45$ giving a

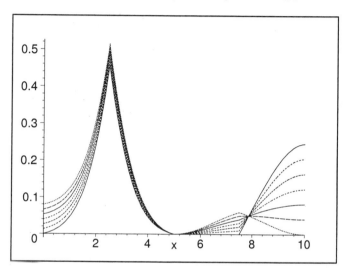

Figure 7.7: Densities generated from $V(NCKB)$ as p_3 ranges from 0.3 to 0.6, assuming a uniform initial prior.

maximally consistent solution with maximum entropy of the form (see figure 7.6, solid line):

$$f(x|KB) = 0.625f(x|s \wedge m \wedge \neg l) + 0.1875f(x|\neg m \wedge l) + 0.1875f(x|\neg m \wedge \neg l)$$

Comparing figures 7.6 and 7.7 we see that the NIS_{KB} density has a significant peak around 7.5 which is not shared by any of the NCS_{KB} densities.

If we are assuming that the expert has attempted to provide a consistent distribution rather than allocating probabilities independently then in the case that there are consistent distributions in $V(KB)$ (i.e. $V(KB) \cap V(CON) \neq \emptyset$) we would expect our inference process to select one of these distributions. The following theorem shows that this is indeed the case for the max. ent. nearest consistent solution to KB.

THEOREM 133 *Consistent KB*
If $V(KB) \cap V(CON) \neq \emptyset$ then $V(NCKB) = V(KB) \cap V(CON)$

Proof
Note that for $\vec{p} \in V(\mathcal{K})$ if $\sum_{i=1}^{t} p_i = 0$ then $\vec{\tilde{p}} = \vec{p}$

$$V(NCKB) = \left\{ \vec{\tilde{p}} : \vec{p} \in V(KB), \sum_{i=t+1}^{2^v} p_i > 0 \text{ and } \sum_{i=1}^{t} p_i = \min_{V(KB)} \sum_{i=1}^{t} p_i \right\}$$

$$= \left\{ \vec{\tilde{p}} : \vec{p} \in V(KB), \sum_{i=1}^{t} p_i = 0 \right\} \text{ since } V(KB) \cap V(CON) \neq \emptyset$$

$$= \left\{ \vec{p} : \vec{p} \in V(KB), \sum_{i=1}^{t} p_i = 0 \right\} = V(KB) \cap V(CON)$$

as required. $\square$

EXAMPLE 134 *Let $KB = \{Pr(small \wedge medium) = 0.7, Pr(large) = 0.2\}$ where $LA = \{small, medium, large\}$ as defined in example 132. Then the atoms of KB are:*

$$P((s \wedge m) \wedge l) = p_1, \ P((s \wedge m) \wedge \neg l) = p_2$$
$$P(\neg(s \wedge m) \wedge l) = p_3, \ P(\neg(s \wedge m) \wedge \neg l) = p_4$$

and

$$V(KB) =$$

$$\left\{ \langle p_1, p_2, p_3, p_4 \rangle \in [0,1]^4 : p_1 + p_2 = 0.7, p_1 + p_3 = 0.2, \sum_{i=1}^{4} p_i = 1 \right\}$$

$$= \{ \langle p_1, 0.7 - p_1, 0.2 - p_1, 0.1 + p_1 \rangle : p_1 \in [0, 0.2] \}$$

Now $\lambda\left((s \wedge m) \wedge l\right) \cap \mathcal{F} = \emptyset$ and hence

$$V(CON) = \left\{ \langle 0, q_2, q_3, q_4 \rangle : \sum_{i=2}^{4} q_1 = 1 \right\}$$

In this case:

$$V(NCKB) = V(KB) \cap V(CON) = \{ \langle 0, 0.7, 0.2, 0.1 \rangle \}$$

so that the nearest consistent density (as shown by the solid line in figure 7.8) is given by:

$$f(x|KB) = 0.7 f(x|s \wedge m \wedge \neg l) + 0.2 f(x|\neg(s \wedge m) \wedge l)$$
$$+ 0.1 f(x|\neg(s \wedge m) \wedge \neg l)$$

This can be compared with the normalised independent solution density (shown by the dashed line in figure 7.8) given by:

$$f(x|KB) = 0.6512f(x|s \wedge m \wedge \neg l) + 0.0698f(x|\neg(s \wedge m) \wedge l)$$
$$+0.2791f(x|\neg(s \wedge m) \wedge \neg l)$$

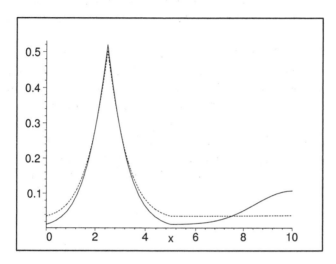

Figure 7.8: Nearest consistent density (solid line) and normalised independent density (dashed line), assuming a uniform initial prior

7.2   Combining Label Expressions with Data

In this section we apply label semantics to the fusion of data and expert knowledge in classification problems and in reliability analysis. Both these application domains involve data modelling in the presence of imprecise background knowledge. In many classical approaches the latter is neglected if not ignored completely and we aim to show that by incorporating the expert knowledge into the data analysis we can significantly enhance the performance of simple data models which in turn can lead to improved generalisation. For example, as discussed in chapter 6, the estimating of probability distributions from data suffers from the so-called 'curse of dimensionality' [8] whereby the amount of data required to accurately estimate a joint probability distribution increases exponentially with the dimension of that distributions. This problem can be avoided if we can limit ourselves to only evaluating marginal distributions for individual attributes and then estimating the joint distribution by assuming independence. Unfortunately, when the independence assumption is not consistent with the data this can lead to decomposition error with spurious probability measure being assigned to regions of the attribute space where there is little or no data. However, such decomposition error can be eliminated if inherent dependencies can be expressed in terms of logical statements which are then fused with estimated distributions, thereby avoiding the necessity of directly evaluating joint distributions prone to the 'curse of dimensionality'. Hence, for both the problems described below we use independent mass relations to model the data and fuse this with background knowledge to enhance performance and reduce decomposition error. In label semantics the fusion of linguistic expression with data can be performed using mass relations as follows:

Suppose that, in addition to the initial prior distribution f on x, we also have background knowledge in the form of a knowledge base KB (as given in definition 121) and database DB. We now propose a new method for fusing expert knowledge and the database with the label semantics framework [68].

DEFINITION 135 *(Fusion in Label Semantics)*
We first evaluate m_{DB} (as described in chapter 4) and then condition on $KB = \{\theta_1, \ldots, \theta_v\}$ (according to definition 94, chapter 5) to give the following mass relation:
$\forall T_i \subseteq LA_i : i = 1, \ldots, k$

$$
m_{DB,KB}(T_1, \ldots, T_k)
$$
$$
= \begin{cases} \dfrac{m_{DB}(T_1,\ldots,T_k)}{\sum_{\langle T_1,\ldots,T_k \rangle \in \lambda^{(n)}(\bigwedge_{i=1}^{v} \theta_i)} m_{DB}(T_1,\ldots,T_k)} & : \langle T_1, \ldots, T_k \rangle \in \lambda^{(k)}(\bigwedge_{i=1}^{v} \theta_i) \\ 0 & : otherwise \end{cases}
$$

We now evaluate the distribution from this mass assignment based on the initial prior so that,

$$
f(\vec{x}|DB, KB) = f(\vec{x}|m_{DB,KB})
$$

according to definition 96 (chapter 5).

This approach can be easily extended to the case where we have an uncertain knowledge base as follows:

DEFINITION 136 *(Fusion given Uncertain Knowledge)*
If KB is an uncertain knowledge base (definition 123) and $\vec{q} \in V(CON)$ is the nearest or most representative consistent solution to the elements of $V(KB)$ then the fusion method is as follows:

$$
\forall T_i \subseteq LA_i : i = 1, \ldots, k \, m_{DB,KB}(T_1, \ldots, T_k) = \sum_{i=t+1}^{2^v} q_i m_{DB,\alpha_i}(T_1, \ldots, T_k)
$$

We then evaluate the distribution from this mass assignment based on the initial prior so that,

$$\forall \vec{x} \in \Omega_1 \times \ldots \times \Omega_k \ f(\vec{x}|DB, KB) = f(\vec{x}|m_{DB,KB}) = \sum_{i=t+1}^{2^v} q_i f(\vec{x}|DB, \alpha_i)$$

7.2.1   Fusion in Classification Problems

Suppose for classification problems as described in chapter 6, we have background knowledge KB_j for each class $c_j \in \mathcal{C}$. Definition 135 can then be used to fuse this knowledge with the mass assignment m_{DB_j} inferred from the database DB_j. To avoid the curse of dimensionality then m_{DB_j} may be evaluated according to an independence or semi-independence assumption as given in definitions 76 and 77 (chapter 4). Accuracy of the mass relational classifier, will tend to be improved by fusion in those cases where the background knowledge encodes dependencies between the attributes that, as a result of the independence assumptions, are not captured by the mass relations.

EXAMPLE 137 *Fusion in the Figure of Eight Problem*
Recall the figure of eight problem as described in example 101 (chapter 6). As in chapter 6 the training database consisted of a regular grid of 961 points on $[-1.6, 1.6]^2$ and the test set of a denser grid of 2116 points. For this example, however, attribute universes Ω_1 and Ω_2 where discretized using 6 labels so that:

$$LA_1 = \{very\ low_1,\ low_1,\ fairly\ low_1,\ fairly\ high_1,\ high_1,\ very\ high_1\}$$
and
$$LA_2 = \{very\ low_2,\ low_2,\ fairly\ low_2,\ fairly\ high_2,\ high_2,\ very\ high_2\}$$

In each case the appropriateness measures were defined as trapezoids (definition 100, chapter 6) generated from a uniform partition of $[-1.6, 1.6]$.

As commented in chapter 6, the figure of eight is a type of XOR problem and hence is prone to decomposition error. This suggests that class models based on independent mass relations are unlikely to give optimal results. For example, figure 7.9 gives a scatter plot of the errors from the independent mass relational model, showing that the dependence between x_1 and x_2 has not been captured. This is further illustrated by the density function for legal (figure 7.10) which does not reflect the actual distribution of legal points.

However, now we show that if such models are fused with suitable expert knowledge then we can significantly improve performance over both the data and expert derived models. In addition, to the database DB we now introduce expert knowledge in the form of the following two linguistic expression

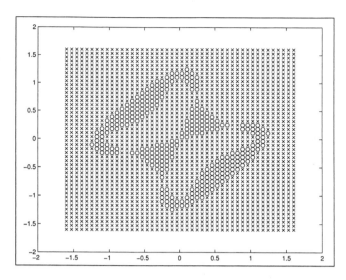

Figure 7.9: Scatter plot of classification results using independent mass relations to model each class. Crosses represent points correctly classified and zero represent points incorrectly classified.

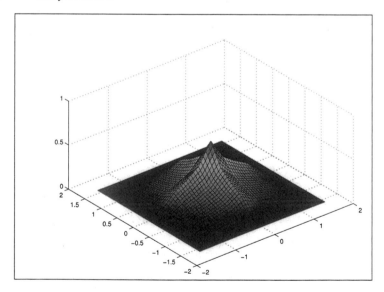

Figure 7.10: Density derived from the independent mass relation for the sub-database of legal elements.

describing the two classes, legal and illegal respectively:

$$KB_1 = \{[(low_1 \vee fairly\ low_1) \wedge (high_2 \vee fairly\ high_2)]$$
$$\vee [(high_1 \vee fairly\ high_1) \wedge (low_2 \vee fairly\ low_2)]\}$$
$$KB_2 = \{\neg [(low_1 \vee fairly\ low_1) \wedge (high_2 \vee fairly\ high_2)]$$
$$\wedge \neg [(high_1 \vee fairly\ high_1) \wedge (low_2 \vee fairly\ low_2)]\}$$

Figure 7.11 shows a scatter plot of the errors obtained using only the knowledge base for classification. In this case the mass relations for legal and illegal are taken as m_{KB_1} and m_{KB_2} respectively. However, this approach overestimates the extent of the legal region, as can be clearly seen from figure 7.12 showing a plot of the legal density as derived from KB_1.

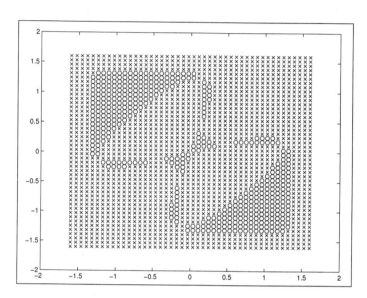

Figure 7.11: Scatter plot of classification results using expert knowledge only to model each class. Crosses represent points correctly classified and zero represent points incorrectly classified.

The models based on the data derived mass relations were then fused with the those based on the background knowledge according to the method proposed in definition 135 to give mass relations m_{DB_1,KB_1} and m_{DB_2,KB_2}. These in turn were used to approximate the class densities so that $f(x, y|legal) \approx f(x_1, x_2|m_{DB_1,KB_1})$ and $f(x_1, x_2|illegal) \approx f(x_1, x_2|m_{DB_2}, KB_2)$ from which bayesian estimates of the class probabilities could then be obtained. The corresponding scatter plot and legal density are shown in figures 7.13 and 7.14 respectively. The percentage accuracy for the 3 models on the training and test sets are shown in figure 7.15.

We now investigate the fusion of uncertain knowledge and data for this problem. The following two uncertain knowledge bases for legal and illegal where devised so that the label expressions covered the whole of the space representing the relevant class. In order to provide meaningful probability values a fully composed mass relation (i.e. without the independence assumption) was inferred for both classes and this was then used to calculate the probability of each label expression by summing over the appropriate λ-set.

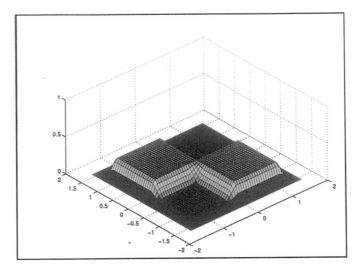

Figure 7.12: Density generated from KB_{legal} for legal.

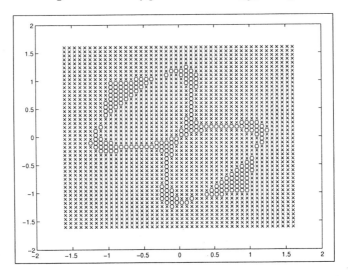

Figure 7.13: Scatter plot of classification results using both background knowledge and independent mass relations to model each class. Crosses represent points correctly classified and zero represent points incorrectly classified.

$KB_1 = \{P\left((low_1 \vee fairly\ low_1) \wedge (high_2 \vee fairly\ high_2)\right) = 0.463,$
$P\left((high_1 \vee fairly\ high_1) \wedge (low_2 \vee fairly\ low_2)\right) = 0.463,$
$P\left(very\ low_1\right) = 0.023, P\left(very\ high_1\right) = 0.023,\ P\left(very\ low_2\right) = 0.023,$
$P\left(very\ high_2\right) = 0.023,\ P\left((low_1 \vee fairly\ low_1) \wedge (low_2 \vee fairly\ low_2)\right)$
$= 0.298,\ P\left((high_1 \vee fairly\ high_1) \wedge (high_2 \vee fairly\ high_2)\right) = 0.298\}$

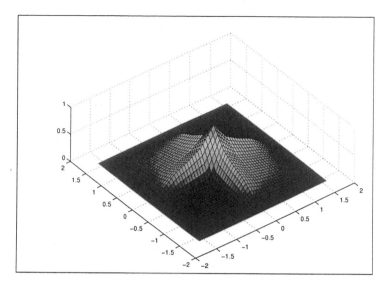

Figure 7.14: Density generated from fused model for legal

	knowledge base and data	data only	knowledge base only
Training error	9.4%	15%	19.1%
Test error	10.6%	14.9%	21.1%

Figure 7.15: Results for figure of eight classification problem

$KB_2 =$

$\{P\left((low_1 \vee fairly\ low_1) \wedge (high_2 \vee fairly\ high_2)\right) = 0.118,$

$P\left((high_1 \vee fairly\ high_1) \wedge (low_2 \vee fairly\ low_2)\right) = 0.118,$

$P\left(very\ low_1\right) = 0.218, Pr\left(very\ high_1\right) = 0.218,$

$P\left(very\ low_2\right) = 0.218, Pr\left(very\ high_2\right) = 0.218,$

$P\left((low_1 \vee fairly\ low_1) \wedge (low_2 \vee fairly\ low_2)\right) = 0.173,$

$P\left((high_1 \vee fairly\ high_1) \wedge (high_2 \vee fairly\ high_2)\right) = 0.173, \}$

Densities were inferred from KB_1 and KB_2 using the normalised independent solutions (definition 128) NIS_{KB_1} and NIS_{KB_2}. These provided estimates of the class densities and from which bayesian classification methods could then be applied. The results for classification based on only the uncertain data are given in column 3 of figure 7.16. The associated scatter plot and density for the legal class are shown in figures 7.17 and 7.18 respectively.

The uncertain knowledge was then fused with the independent mass relations described above using the method outlined in definition 136. The results for

	uncertain knowledge and data	data only	uncertain knowledge only
Training error	7.1%	15%	23.1%
Test error	7.4%	14.9%	23.3%

Figure 7.16: Figure of eight classification results based on uncertain knowledge

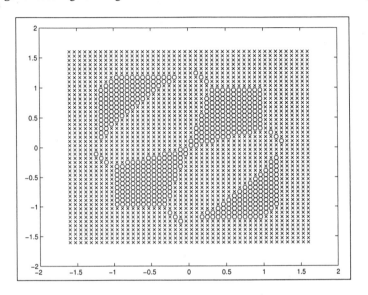

Figure 7.17: Scatter plot of classification results using uncertain expert knowledge only to model each class. Crosses represent points correctly classified and zero represent points incorrectly classified.

classification based on this fused model are given in column 1 of the table in figure 7.16. The associated scatter plot and density for the legal class are shown in figures 7.19 and 7.20 respectively.

7.2.2   Reliability Analysis

Engineering reliability analysis is conventionally based on the use of proba-bilistic information about the loads and responses of an engineering system to estimate the system probability of failure. Whilst use of reliability methods is now widespread, they have been criticised on several grounds [11], [29]. These include the constraints that the information input into the analysis has to be in a precise probabilistic format and the surface that divides 'failed' and 'not failed' system states (known as the limit state function) is precisely known. The former constraint has been addressed quite widely by reformulating reliability calculations to accept information in a range of formats, including probability intervals [17], fuzzy sets [11], [14], [16], convex modelling and random sets [100]. The latter problem of the precise form of the limit state function is more

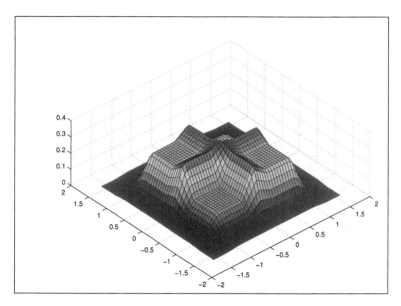

Figure 7.18: Density generated from uncertain knowledge base for legal.

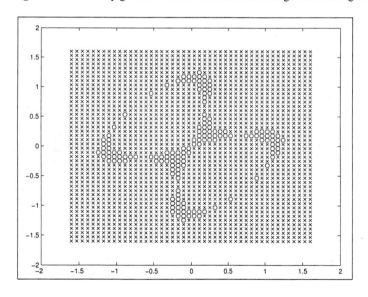

Figure 7.19: Scatter plot of classification results using the fused model from uncertain expert
knowledge and data to model each class. Crosses represent points correctly classified and zeros represent points incorrectly classified.

profound and difficult to address. Conventionally, uncertainty in the limit state
function has been model by adding another random variable to the state variable
set to represent uncertainty. However, the empirical meaning of this variable is
far from clear and its precise probabilistic format can be hard to justify.

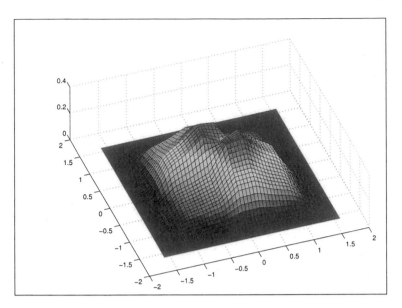

Figure 7.20: Density generated from uncertain knowledge base fused with data for legal.

Indeed in many cases where there is significant uncertainty regarding the limit state function there is imprecise information regarding the system behaviour in the form of expert knowledge. There may also be some data available resulting from the experimental testing of the system to determine parameter values at which failure occurs or as examples of past failures. However, such testing is likely to be expensive and past failure of high reliability systems are, by definition, rare so that data tends to be sparse. Given the 'curse of dimensionality' this means that it may be difficult to estimate joint distributions on the parameter space and that the use of decomposed (independent) models may be necessary. In this context the use of available background knowledge could be essential in order to limit decomposition error.

More formally, reliability analysis calculates P_f, the probability of failure of a system characterised by a vector $\vec{x} = \langle x_1, \ldots, x_n \rangle$ of basic variables on $\mathbb{R}^n$. The limit state function $g(\vec{x})$ represents the difference between the resistance of the system and the loading effect given physical parameters $\vec{x}$. The probability of failure P_f is the probability that loading exceeds resistance so that:

$$P_f = P(g(\vec{x}) \leq 0)$$

It is usually assumed that $g(\vec{x})$ is monotonic and therefore, given prior knowledge of the distribution of $\vec{x}$, P_f can be calculated provided we can identify the set of parameter values for which $g(\vec{x}) = 0$. In effect this corresponds to the requirement that we can precisely partition the parameter space into two distinct regions corresponding to 'not failed' and 'failed' states of the system.

In the following example, we consider the case were the actual knowledge is in fact only sufficient to divide the space into fuzzy regions for 'not failed', 'failed' and 'doubtful'.

EXAMPLE 138 *(Structural Reliability of Coastal Defences)*
This example is based on a previous conventional reliability analysis of a dike on the Frisian coast in the Netherlands, along the Wadden Sea [13]. An initial attempt to model this problem using label expressions can be found in [43]. The analysis presented here is based on that given in [68]. The behaviour of the concrete block revetment on the seaward slope of the dike is described by the basic variables $\vec{x} = \langle \Delta, D, H_s, \alpha, s_{op} \rangle$ where:

Δ *is the density of the revetment blocks*

D *is the diameter of the revetment blocks*

H_s *is the significant wave height*

α *is the slope of the revetment*

s_{op} *is the offshore wave steepness*

To reduce the dimensionality of the problem these variables may be arranged as two non-dimensional groups: $\frac{H_s}{\Delta D}$ and $\xi_{op} = s_{op}^{-0.5} \tan \alpha$. This parameter space would conventionally be divided into 'failed' and 'not failed' regions. However, there is very little physical justification for such a strict definition of the limit state function and in fact background knowledge of the problem can be better described in terms of an imprecise (doubtful) region as shown in figure 7.21 where it is possible that $g(\vec{x}) = 0$. For this example, a fuzzy version of this region was defined on the basis of the label sets $LA_1 = \{very\ low_1, low_1, medium_1, high_1, very\ high_1\}$ and $LA_2 = \{very\ low_2, low_2, medium_2, high_2, very\ high_2\}$ for ξ_{op} and $\frac{H_s}{\Delta D}$ respectively. In both cases the labels were defined according to trapezoidal appropriateness measures based on uniform partitions of the universes $[0, 6]$ and $[0, 9]$ respectively. The doubtful region was then described by the following label expression:

$doubtful \equiv$
$[(very\ low_1 \wedge \neg\ low_1) \wedge (\neg low_2 \wedge (medium_2 \vee high_2 \vee very\ high_2))]$
$\vee\ [(very\ low_1 \wedge low_1) \wedge ((low_2 \wedge \neg very\ low_2) \vee (medium_2 \wedge \neg high_2))]$
$\vee\ [(low_1 \wedge \neg very\ low_1 \wedge \neg medium_1) \wedge (low_2 \wedge \neg medium_2)]$
$\vee\ [(low_1 \wedge medium_1) \wedge (low_2 \wedge \neg very\ low_2)]$
$\vee\ [(medium_1 \wedge \neg low_1) \wedge (low_2 \wedge very\ low_2)]$
$\vee\ [((high_1 \wedge \neg medium_1) \vee very\ high_2) \wedge (low_2 \wedge \neg medium_2)]$

The two dimension appropriateness measure for this expression, as described in chapter 4, is shown in figure 7.22 as a contour plot. If we assume a uniform prior distribution then the prior density for the limit state function (i.e. the distribution of values of ξ_{op} and $\frac{H_s}{\Delta D}$ for which $g(\vec{x}) = 0$) will be proportional to this appropriateness measure.

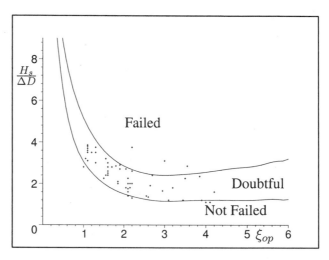

Figure 7.21: Classification of the parameter space in condition assessment guidance for flood defence revetments [13]

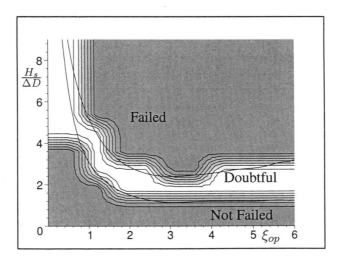

Figure 7.22: Contour plot showing the label based classification of the parameter space.

A small number of data values (also shown in figure 7.21) are available for this dike revetment. Each point represents measurements at failure in experimental tests. The data is expensive to obtain and hence there are only

*a small number of values. An independent mass relation was obtained from
this data for the above label sets and assuming a uniform prior distribution on
$[0, 6] \times [0, 9]$ a density function (definition 96, chapter 5) was then obtained
(see figure 7.23). The resulting distribution shows clear signs of decomposition
error resulting from an inappropriate use of the independence assumption. For
example, non-zero density is given to points in the region $[1, 1.6] \times [0, 0.8]$ which
is not supported by the data.*

*Fusing the independent mass relation with background knowledge in the form
of the above label description of the doubtful region results in a posterior density
as shown in figure 7.24. Here the effects of decomposition error have been
greatly reduced and the resulting distribution conforms to the expert judgement
as is to be expected. In certain situations it is possible that experts will*

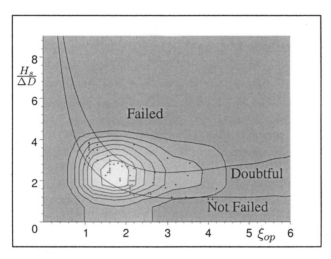

Figure 7.23: Contour plot showing the density derived from the independent mass relation and based on the data only.

*have some prior knowledge regarding the likely distribution of points across
different regions of the doubtful area. To simulate this for the current example
four fuzzy regions (see figure 7.25) were defined according to the following label
expressions:*

$region1 \equiv$
$[(very\ low_1 \wedge \neg low_1 \wedge \neg medium_1)$
$\wedge ((very\ high_2 \vee high_2 \vee medium_2) \wedge \neg low_2)]$
$\vee [(very\ low_1 \wedge low_1) \wedge ((medium_2 \wedge \neg high_2) \vee (low_2 \wedge \neg very\ low_2))]$
$\vee [(low_1 \wedge \neg very\ low_1)$
$\wedge ((low_2 \wedge \neg medium_2 \wedge \neg very\ low_2) \vee (very\ low_2 \wedge \neg low_2))]$

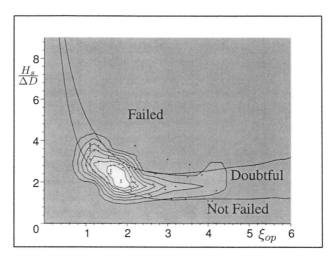

Figure 7.24: Contour plot showing the density derived from the independent mass relation fused with the fuzzy classification of the doubtful region.

$region2 \equiv$

$[(medium_1 \wedge high_1) \wedge (very\ low_2 \wedge low_2)]$

$\vee [(very\ high_1) \wedge (low_2 \wedge \neg very\ low_2 \wedge \neg medium_2)]$

$\vee [(very\ high_1 \wedge \neg high_1) \wedge (very\ low_2 \wedge low_2)]$

$region3 \equiv$

$[(medium_1 \vee high_1 \vee very\ high_1) \wedge (very\ low_2 \wedge \neg low_2)]$

$\vee [(very\ low_1 \wedge low_1) \wedge (verylow_2 \wedge low_2)]$

$\vee [(very\ low_1 \wedge \neg low_1) \wedge (low_2 \wedge medium_2)]$

$region4 \equiv$

$[((medium_1 \vee high_1) \wedge \neg very\ high_1) \wedge (very\ low_2 \wedge \neg low_2)]$

$\vee [(low_1 \wedge \neg very\ low_1) \wedge (low_2 \wedge medium_2)]$

$\vee [(very\ low_1 \wedge \neg low_1) \wedge (medium_2 \wedge high_2)]$

Probabilities for the regions were then evaluated according to:

$$P(region) = \frac{\int_{[0,6]} \int_{[0,9]} \mu_{region}(x,y)\ doubtful(x,y)dxdy}{\int_{[0,6]} \int_{[0,9]} doubtful(x,y)dxdy}$$

where $doubtful(x,y)$ *denotes the characteristic function value of the original crisp doubtful area as shown in figure 7.21. Hence, the probability of each*

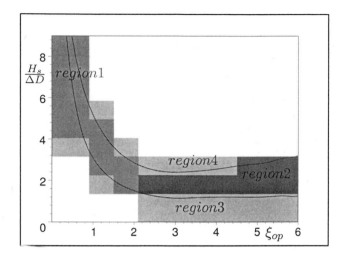

Figure 7.25: Regions partitioning the doubtful area based on label expression

region corresponds to the conditional probability of the corresponding label expression given that (x, y) lies within the doubtful region. This gives:

$$KB = \{P(region1) = 0.29, P(region2) = 0.515, P(region3) = 0.095,$$
$$P(region4) = 0.1\}$$

In practice, it is envisaged that such probabilities would be provided by domain experts. In this case, the label description of the regions are mutually exclusive and exhaustive so that applying the fusion of uncertain knowledge method given in definition 136 based on the max. ent. nearest consistent solution to KB (definition 131) gives:

$$\forall x \in [0, 6], \forall y \in [0, 9] \ f(x, y|KB) =$$
$$0.29 f(x, y|region1) + 0.515 f(x, y|region2) +$$
$$0.095 f(x, y|region3) + 0.1 f(x, y|region4)$$

The resulting density, shown in figure 7.26, is more skewed towards region2 than would be suggested by the data alone because of the high prior probability given to that region. This can be seen clearly when comparing the contour plots in figures 7.24 and 7.26.

Summary

In this chapter we have investigated using the label semantics framework to fuse background knowledge, in the form of label expressions, and data. In particular, we have shown how the results and definitions of chapter 5 on conditional probability given label expressions can be applied to generate informative

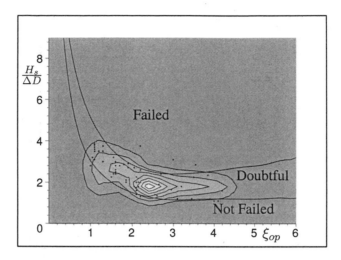

Figure 7.26: Contour plot showing the density derived from the independent mass relation fused with uncertain description of the doubtful region.

priors from qualitative background knowledge. This approach was extended to allow for the case where the prior knowledge is uncertain and hence corresponds to a set of subjective probability values on expressions. In this context the atoms of a knowledge base identify all possible conjunctions of the label expressions in that knowledge base and their negations. We then identified two inference processes for selecting a unique probability distribution on atoms, which is non-zero only on those atoms consistent with the label definitions. In the case of the normalised independent solution (NIS_{KB}) the underlying assumption is that during the knowledge elicitation process, the expert took no account of the interpretation of the labels and how they inter-relate. Instead, he or she evaluated the probability of each expression, more or less independently of all the other expressions. Alternatively, the nearest consistent solution ($NCKB$) assumes that the expert attempts to be consistent in the allocation of their probability values and that any inconsistency arises as a result of an error in this process. Given this assumption it is natural to identify the nearest consistent probability distributions on atoms, where distance between probability distributions is measures using cross (relative) entropy. In the case that this does not determine a unique distribution on atoms, maximum entropy is used to identify the distribution with minimal extra information, beyond that imposed by the constraints in the knowledge base. Once a unique distribution has been selected on the consistent atoms then the informative prior distribution is taken to be the linear combination, relative to these probabilities values, of the conditional distributions given each atom as evaluated from an uninformative prior.

In the second part of the chapter we investigated the application of label semantics based fusion to problems in classification, and in reliability analysis. For these problems fusion takes place at the level of mass relations using the formula for conditional mass relations given in chapter 5. For classification problems label expressions can be used to describe known dependencies between attributes enabling independent mass relations to better represent the data. In reliability analysis label expressions can represent qualitative prior knowledge about the limit state function, which can compensate for sparsity of data when estimating probabilities of system failure. To illustrate these ideas we presented an example of reliability analysis of a dike on the Frisian coast of the Netherlands which showed how label expressions can be used to capture linguistic information about the limit state function.

Chapter 8

NON-ADDITIVE
APPROPRIATENESS MEASURES

In this final chapter we return once again to consider foundational issues regarding the label semantics calculus. In particular, we shall focus on relaxing the additive condition, assumed implicity throughout the volume and stated explicitly in axiom AM4 (chapter 3). Here we shall weaken this condition to allow for the possibility that Your beliefs concerning the appropriateness of labels are aggregated in a non-additive manner using a t-conorm. In order to express this idea more formally it is first necessary to give a generalised definition of mass assignments on labels:

DEFINITION 139 *Generalised Mass Assignment*
A generalised mass assignment on labels is a function $m : 2^{LA} \to [0,1]$

Effectively this is a simple generalisation of definition 24 (chapter 3) obtained by removing the condition that the sum of the masses across all label sets is one. As before $m_x(T)$ for $T \subseteq LA$ is interpreted as Your belief that T corresponds to the entire set of labels with which it is appropriate to describe x (i.e. that $\mathcal{D}_x = T$). The difference will now lie, however, in the way in which mass values are aggregated in order to evaluate appropriateness measures of label expressions. Now since $\lambda(\theta)$ identifies the set of appropriate label sets consistent with expression $\theta \in LE$ then it would seem natural for You to evaluate the appropriateness of θ for describing x as a disjunctive combination of the mass values of those sets in $\lambda(\theta)$. This motivates the following generalisation of the definition of appropriateness measure:

DEFINITION 140 *Generalised Appropriateness Measures*

$$\forall \theta \in LE, \ \forall x \in \Omega \ \mu_\theta(x) = f_\vee(m_x(T) : T \in \lambda(\theta))$$

where $f_\vee$ *is a t-conorm as defined in chapter 2*

Notice that this is a strict generalisation of the appropriateness measure calculus introduced in chapter 3 which can be obtained by taking $f_\vee(a,b) = \min(1, a+b)$ and requiring that $\sum_{T \subseteq LA} m_x(T) = 1$.

To obtain a functional calculus then, as with additive appropriateness measures, we must introduce a mapping from appropriateness measures of labels (i.e. $\mu_L(x) : L \in LA$) to generalised mass assignments. Such a mapping extends definition 35 (chapter 3) and is referred to as generalised mass selection function (gmsf):

DEFINITION 141 *Generalised Mass Selection Function*
Let $\mathcal{GM}$ be the set of all generalised mass assignments on 2^{LA}. Then a generalised mass selection function is a function $\Delta : [0,1]^n \to \mathcal{GM}$ such that if $\forall x \in \Omega\ \Delta\left(\mu_{L_1}(x), \dots, \mu_{L_n}(x)\right) = m_x$ then

$$\forall L \in LA\ f_\vee\left(m_x(T) : L \in T\right) = \mu_L(x)$$

We note that for every generalised appropriateness measure there is at least one valid generalised mass selection function since for every t-conorm $f_\vee$ the following is a valid (if trivial) gmsf:

$$\forall T \subseteq LA : |T| \neq 1\ m_x(T) = 0 \text{ and } \forall L \in LA\ m_x(\{L\}) = \mu_L(x)$$

In the following section we investigate a number of properties of generalised appropriateness measures and show how generalised mass selection functions can be linked to t-norms.

8.1 Properties of Generalised Appropriateness Measures

We now present some results on generalised appropriateness measures, largely relating to the case where $f_\vee$ is either an Archimedean or strict Archimedean t-conorm. We also introduce a property whereby the appropriateness measure of conjunctions of labels is evaluated by application of a t-norm to the appropriateness of each conjunct.

DEFINITION 142 *(Archimedean t-conorms)*
$f_\vee : [0,1]^2 \to [0,1]$ is an Archimedean t-conorm if $f_\vee$ is a t-conorm and also satisfies:

- *$f_\vee$ is continuous*

- *$\forall a \in (0,1)\ f_\vee(a,a) > a$*

DEFINITION 143 *(Strict Archimedean t-conorms)*
$f_\vee : [0,1]^2 \to [0,1]$ is a strict Archimedean t-conorm if $f_\vee$ is an Archimedean t-conorm and is strictly increasing on $(0,1)$.

Archimedean t-conorms can be generated according to an additive generation function g as can be seen from the following theorem (see [53]):

THEOREM 144 *A function* $f_\vee : [0,1] \times [0,1] \rightarrow [0,1]$ *is an Archimedean t-conorm iff there exists an strictly increasing and continuous function* $g : [0,1] \rightarrow [0,\infty]$ *with* $g(0) = 0$ *such that*

$$\vee_t(a,b) = g^{(-1)}(g(a) + g(b))$$

where $g^{(-1)}$ *is the pseudoinverse of* g*, defined by:*

$$g^{(-1)}(y) = \begin{cases} g^{-1}(y) & : & y \in [0, g(1)] \\ 1 & : & x \in (g(1), \infty] \end{cases}$$

Moreover, $f_\vee$ *is strict iff* $g(1) = \infty$ *(i.e. in this case* $\forall y \in [0,\infty]$ $g^{(-1)}(y) = g^{-1}(y)$*)*
Proof *See [105] or [53]* $\square$

DEFINITION 145 *([105])*

$$\forall a, b \in [0,1] \; a \dot{-} b = \inf\{\alpha : f_\vee(b, \alpha) \geq a\}$$

LEMMA 146 *If* $f_\vee$ *is an Archimedean t-conorm then* $\forall x \in \Omega$

$$\forall A \subseteq B \subseteq 2^{LA} \; M_x(B - A) \geq M_x(B) \dot{-} M_x(A)$$

where for $A \subseteq 2^{LA}$ $M_x(A) = f_\vee(T : T \in A)$
Proof *Special case of a results presented in [105].* $\square$

The following theorem gives some indication as to the behaviour of negation for generalised appropriateness measures. Notice that for generalised appropriateness measures it is not necessarily the case that $\mu_{\theta \wedge \neg\theta}(x) = 1$, a fact that we will discuss in more detail in a later section.

THEOREM 147 *If* $f_\vee$ *is an Archimedean t-conorm then*

$$\forall x \in \Omega \; \forall \theta \in LE \; \mu_{\neg\theta}(x) \geq \mu_{\theta \vee \neg\theta}(x) \dot{-} \mu_\theta(x)$$

Proof

$$\forall \theta \in LE \; \lambda(\theta \vee \neg\theta) = 2^{2^{LA}} \Rightarrow \lambda(\theta) \subseteq \lambda(\theta \vee \neg\theta)$$

Also

$$\lambda(\neg\theta) = \lambda(\theta)^c = 2^{2^{LA}} - \lambda(\theta) = \lambda(\theta \vee \neg\theta) - \lambda(\theta)$$

Therefore, by lemma 146 $\forall x \in \Omega$

$$M_x\left(\lambda(\neg\theta)\right) \geq M_x\left(\lambda(\theta \vee \neg\theta)\right) \dot{-} M_x\left(\lambda(\theta)\right) \Rightarrow \mu_{\neg\theta}(x) \geq \mu_{\theta\vee\neg\theta}(x) \dot{-} \mu_\theta(x)$$

as required. $\square$

Note that if $f_\vee$ is a strict Archimedean t-conorm and $M_x(A) < 1$ then the inequality in lemma 146 can be replaced with equality (see [105]). Hence, in this case theorem 147 can be modified so that

$$\mu_{\neg\theta}(x) < 1 \Rightarrow \mu_{\neg\theta}(x) = \mu_{\theta\vee\neg\theta}(x) \dot{-} \mu_\theta(x)$$

LEMMA 148

$$\forall x \in \Omega \; \forall T_1, T_2 \subseteq 2^{LA}$$
$$f_\vee\left(M_x\left(T_1 \cup T_2\right), M_x\left(T_1 \cap T_2\right)\right) = f_\vee\left(M_x\left(T_1\right), M_x\left(T_2\right)\right)$$

Proof *Special case of a result presented in [105].* $\square$

THEOREM 149 *If μ is a generalised appropriateness measure based on strict Archimedean t-conorm $f_\vee$ with generator function g then*

$$\forall x \in \Omega \; \forall \theta, \varphi \in LE \; \mu_{\theta\vee\varphi}(x) = g^{-1}\left(g\left(\mu_\theta(x)\right) + g\left(\mu_\varphi(x)\right) - g\left(\mu_{\theta\wedge\varphi}(x)\right)\right)$$

Proof
By lemma 148 we have that $\forall x \in \Omega \; \forall \theta, \varphi \in LE$

$$\vee_t\left(M_x\left(\lambda(\theta) \cup \lambda(\varphi)\right), M_x\left(\lambda(\theta) \cap \lambda(\varphi)\right)\right) = \vee_t\left(M_x\left(\lambda(\theta)\right), M_x\left(\lambda(\varphi)\right)\right)$$
$$\Rightarrow \vee_t\left(M_x\left(\lambda(\theta \vee \varphi)\right), M_x\left(\lambda(\theta \wedge \varphi)\right)\right) = \vee_t\left(M_x\left(\lambda(\theta)\right), M_x\left(\lambda(\varphi)\right)\right)$$
$$\Rightarrow \vee_t\left(\mu_{\theta\vee\varphi}(x), \mu_{\theta\wedge\varphi}(x)\right) = \vee_t\left(\mu_\theta(x), \mu_\varphi(x)\right)$$

Therefore, since $\vee_t$ is a strict Archimedean t-conorm then by theorem 144

$$g^{-1}\left(g\left(\mu_{\theta\vee\varphi}(x)\right) + g\left(\mu_{\theta\wedge\varphi}(x)\right)\right) = g^{-1}\left(g\left(\mu_\theta(x)\right) + g\left(\mu_\varphi(x)\right)\right)$$

Applying g to both sides gives

$$g\left(\mu_{\theta\vee\varphi}(x)\right) = g\left(\mu_\theta(x)\right) + g\left(\mu_\varphi(x)\right) - g\left(\mu_{\theta\wedge\varphi}(x)\right)$$
$$\Rightarrow \mu_{\theta\vee\varphi}(x) = g^{-1}\left(g\left(\mu_\theta(x)\right) + g\left(\mu_\varphi(x)\right) - g\left(\mu_{\theta\wedge\varphi}(x)\right)\right)$$

as required. $\square$

As discussed in chapter 3 it is interesting to consider the relationship between generalised appropriateness measures and particular t-norms modelling conjunction. As before, we focus on understanding what conditions are imposed on gmsf if we require that the appropriateness measure of a conjunction of basic labels (i.e. members of LA) is given by a t-norm applied to the appropriateness measures of the conjuncts. i.e. when:

$$\forall x \in \Omega, \forall T \subseteq LA : T \neq \emptyset \; \mu_{\left(\bigwedge_{L \in S} L\right)}(x) = f_\wedge \left(\mu_L(x) : L \in T\right)$$

In the sequel we shall refer to this condition as the conjunctive property (**CP**)

LEMMA 150 *If $f_\vee$ is a strict Archimedean t-conorm with generator function g then*

$$\forall u, y, z \in [0, 1] \; \vee_t (y, u) = z \Rightarrow y = g^{-1}\left(g(z) - g(u)\right)$$

Proof
$\forall u, y, z \in [0, 1]$

$$\vee_t (y, u) = z \Rightarrow g^{-1}\left(g(u) + g(y)\right) = z$$
$$\Rightarrow g(u) + g(y) = g(z) \Rightarrow g(y) = g(z) - g(u) \Rightarrow y = g^{-1}\left(g(z) - g(u)\right)$$

as required. $\square$

THEOREM 151 *Let $f_\vee$ be a strict Archimedean t-conorm with generator function g and let μ be a generalised appropriateness measure based on $f_\vee$. If μ satisfies* **CP** *for t-norm $f_\wedge$ then*

$$\forall x \in \Omega \; \forall T \subseteq LA : T \neq \emptyset$$
$$m_x(T) = g^{-1}\left(g\left(f_\wedge\left(\mu_L(x) : L \in T\right)\right) - g\left(f_\vee\left(m_x(T') : T' \supset T\right)\right)\right)$$

Proof
By **CP** *we have that $\forall T \subseteq LA : T \neq \emptyset$*

$$f_\vee\left(m_x(T') : T' \supseteq T\right) = f_\vee\left(m_x(T), f_\vee\left(m_x(T') : T' \supset T\right)\right)$$
$$= f_\wedge\left(\mu_L(x) : L \in T\right)$$

Therefore, by lemma 150 letting $y = m_x(T)$, $u = f_\vee\left(m_x(T') : T' \supset T\right)$ and $z = f_\wedge\left(\mu_L(x) : L \in T\right)$ we obtain

$$m_x(T) = g^{-1}\left(g\left(f_\wedge\left(\mu_L(x) : L \in T\right)\right) - g\left(f_\vee\left(m_x(T') : T' \supset T\right)\right)\right)$$

as required. $\square$

For a fixed $f_\wedge$ the formula given in theorem 151 can be applied recursively to obtain a unique solution for $m_x(T)$ for all $T \subseteq LA$. If this process results in negative mass being allocated to any subset of labels then no gmsf exists that satisfies **CP** for that particular t-norm. Consequently for appropriateness measure based on strict Archimedean t-conorms there is at most one mass selection function satisfying **CP** for any t-norm.

8.2   Possibilistic Appropriateness Measures

As discussed in chapter 5, possibility theory has been strongly linked to fuzzy set theory by the fact that a fuzzy set naturally defines a possibility distribution on the underlying universe Ω. It is also now a well developed theory of uncertainty modelling that is especially useful when information is ordinal or qualitative in nature [26]. In this section we examine the role of possibility theory in label semantics by defining possibilistic appropriateness measures based on the max t-conorm.

DEFINITION 152 *(Possibilistic Appropriateness Measures)*
A Possibilistic appropriateness measure is a generalised appropriateness measure based on t-conorm $f_\vee = \max$ so that:

$$\forall x \in \Omega, \ \forall \theta \in LE, \ \mu_\theta(x) = \max(m_x(T) : T \in \lambda(\theta))$$

Notice that since max is an idempotent t-conorm it holds that

$$\forall \theta, \varphi \in LE \ \forall x \in \Omega \ \mu_{\theta \vee \varphi}(x) = \max(\mu_\theta(x), \mu_\varphi(x))$$

Consequently, for a fixed x, μ is a non-normalized possibility measure on LE. For, μ to correspond to a normalized possibility measure requires the additional condition that:

$$\forall x \in \Omega \ \max(m_x(T) : T \subseteq LA) = 1$$

We now consider the effect of **CP** on possibilistic appropriateness measures. For clarity we shall refer to generalized mass selection functions based on the max t-conorm as possibilistic mass section functions. We now introduce a family of possibilistic msf satisfying **CP** for any choice of t-norms:

DEFINITION 153 *(Conjunctive Possibilistic msf)*
A conjunctive possibilistic msf based on t-norm $f_\wedge$ is defined as follows: Given appropriateness degrees $\mu_{L_1}(x), \ldots, \mu_{L_n}(x)$ then the following generalised mass assignment is selected.

$$\forall T \subseteq LA : T \neq \emptyset$$
$$m_x(T) = f_\wedge(\mu_L(x) : L \in T) \ and \ m_x(\emptyset) = \epsilon \ where \ \epsilon \in [0, 1]$$

In the context of possibilistic appropriateness measures the conjunctive msf has a clear justification as follows: If $T \neq \emptyset$ corresponds to the true set of appropriate labels for x (i.e. $\mathcal{D}_x = T$), then the conjunction of statements of the form 'L is an appropriate label for x' for $L \in T$ must be true. Therefore, to evaluate Your belief that $\mathcal{D}_x = T$ it would seem intuitively reasonable for You combine your beliefs that 'L is an appropriate label for x' (i.e. $\mu_L(x)$)

conjunctively across those labels L contained in T. The value for $m_x(\emptyset)$ is then allocated to correspond to Your belief that x cannot be labelled.

THEOREM 154 *If μ is a possibilistic appropriateness measure defined according to a conjunctive possibilistic msf then*

$$\forall x \in \Omega \; \forall T \subseteq LA \; \mu_{(\wedge_{L \in T} L)}(x) = f_\wedge (\mu_L(x) : L \in T)$$

Proof

$$\forall x \in \Omega \; \forall T \subseteq LA \; \mu_{(\wedge_{L \in T} L)}(x) = \max \left(m_x(T') : T \subseteq T' \right) =$$
$$\max \left(f_\wedge \left(\mu_L(x) : L \in T' \right) : T \subseteq T' \right)$$

Now by the monotonicity of $f_\wedge$ it follows that if $T \subseteq T'$ then
$f_\wedge (\mu_L(x) : L \in T) \geq f_\wedge (\mu_L(x) : L \in T')$ *and therefore*

$$\max \left(f_\wedge \left(\mu_L(x) : L \in T' \right) : T \subseteq T' \right) = f_\wedge (\mu_L(x) : L \in T)$$

as required. □

Note that the conjunctive possibilistic msf for t-norm $f_\wedge$ is not the only possibilistic msf that satisfies **CP** for $f_\wedge$. To see this consider the following example:

EXAMPLE 155 *Let $LA = \{L_1, L_2, L_3\}$ and suppose for some $x \in \Omega$ that $\mu_{L_1}(x) = 0.3$, $\mu_{L_2}(x) = 0.5$ and $\mu_{L_3}(x) = 0.8$. Furthermore, suppose that $f_\wedge = \min$. In this case the conjunctive possibilistic msf selects the following generalized mass assignment:*

$m_x (\{L_1, L_2, L_3\}) = 0.3, m_x (\{L_1, L_2\}) = 0.3,$
$m_x (\{L_1, L_3\}) = 0.3, m_x (\{L_2, L_3\}) = 0.5,$
$m_x (\{L_1\}) = 0.3, m_x (\{L_2\}) = 0.5,$
$m_x (\{L_3\}) = 0.8, m_x (\emptyset) = \epsilon$

*However, an alternative possibilistic msf might select the following generalized mass assignment and still remain consistent with **CP** for min:*

$m_x (\{L_1, L_2, L_3\}) = 0.3, m_x (\{L_1, L_2\}) = 0.1,$
$m_x (\{L_1, L_3\}) = 0.2, m_x (\{L_2, L_3\}) = 0.5,$
$m_x (\{L_1\}) = 0.05, m_x (\{L_2\}) = 0.05,$
$m_x (\{L_3\}) = 0.8, m_x (\emptyset) = \epsilon$

In fact we can show that for a given t-norm $f_\wedge$, the corresponding conjunctive possibilistic msf (defined uniquely up to mass on the empty set) is the maximal msf satisfying **CP** for that t-norm.

THEOREM 156 *Let Δ^c denote the conjunctive possibilistic msf based on t-norm $f_\wedge$. Let Δ be another possibilistic msf satisfying **CP** for the same t-norm $f_\wedge$. If $\forall x \in \Omega\ \Delta^c\left(\mu_{L_1}(x), \ldots, \mu_{L_n}(x)\right) = m_x^c$ and $\forall x \in \Omega\ \Delta\left(\mu_{L_1}(x), \ldots, \mu_{L_n}(x)\right) = m_x$ then*

$$\forall T \subseteq LA : T \neq \emptyset\ m_x\left(T\right) \leq m_x^c\left(T\right) = \wedge_t\left(\mu_L(x) : L \in T\right)$$

Proof
*Since Δ satisfies **CP** for $f_\wedge$ then*

$$\forall x \in \Omega\ \mu_{L_1 \wedge \ldots \wedge L_n}(x) = m_x\left(\{L_1, \ldots, L_n\}\right) = f_\wedge\left(\mu_{L_1}(x), \ldots, \mu_{L_n}(x)\right)$$

We now prove the result by recursion on $|T|$ for $T \subseteq LA$. Suppose, as the recursive hypothesis, that:

$$\forall T : |T| \geq k\ m_x\left(T\right) \leq f_\wedge\left(\mu_L(x) : L \in T\right)$$

*Now consider $T \subseteq LA$ such that $|T| = k - 1 > 0$ then since Δ satisfies **CP** for $f_\wedge$ we have that:*

$$\mu_{\left(\wedge_{L \in T} L\right)}\left(x\right) = \max\left(m_x\left(T\right), \max_{T' \supset T} m_x\left(T'\right)\right)$$

Now by the inductive hypothesis it holds that:

$$\forall T' \supset T\ m_x\left(T'\right) \leq f_\wedge\left(\mu_L\left(x\right) : L \in T'\right)$$

Therefore,

$$\max_{T' \supset T} m_x\left(T'\right) \leq \max_{T' \supset T} f_\wedge\left(\mu_L(x) : L \in T'\right) \leq f_\wedge\left(\mu_L(x) : L \in T\right)$$

by the monotonicity property of t-norms.
Hence, there are two possible cases:

*(i) $m_x(T) = f_\wedge\left(\mu_L(x) : L \in T\right)$ and
    $\max_{T' \supset T} m_x\left(T'\right) \leq f_\wedge\left(\mu_L(x) : L \in T\right)$ or*

*(ii) $\max_{T' \supset T} m_x\left(T'\right) = f_\wedge\left(\mu_L(x) : L \in T\right)$ and
    $m_x(T) \leq f_\wedge\left(\mu_L(x) : L \in T\right)$*

In either case $m_x(T) \leq f_\wedge\left(\mu_L(x) : L \in T\right)$ as required. □

The following theorem gives some insight into the behaviour of possibilistic appropriateness measures on tautologies and on negations of labels:

THEOREM 157 *If μ is a possibilistic appropriateness measure defined according to a conjunctive possibilistic msf then*

(i) $\forall \theta \in LE \ \forall x \in \Omega \ \mu_{\theta \vee \neg \theta}(x) = \max(\epsilon, \mu_{L_1}(x), \dots, \mu_{L_n}(x))$

(ii) $\forall L_i \in LA \ \forall x \in \Omega \ \mu_{\neg L_i}(x) = \max(\epsilon, \max(\mu_L(x) : L \neq L_i))$

Proof
(i)

$\forall x \in \Omega \ \forall \theta \in LE \ \mu_{\theta \vee \neg \theta}(x) = \max \left(m_x(T) : T \subseteq LA \right) =$
$\max \left(m_x(\emptyset), m_x(\{L_1\}), \dots, m_x(\{L_n\}), \max \left(m_x(T) : T \subseteq LA, |T| > 1 \right) \right)$
$= \max \left(\epsilon, \mu_{L_1}(x), \dots, \mu_{L_n}(x), \max \left(f_\wedge \left(\mu_L(x) : L \in T \right) : T \subseteq LA, |T| > 1 \right) \right)$

Now by the monotonicity of $f_\wedge$ it follows that

$\forall L \in T \ f_\wedge \left(\mu_L(x) : L \in T \right) \leq \mu_L(x)$

Hence the above simplifies to

$\max(\epsilon, \mu_{L_1}(x), \dots, \mu_{L_n}(x))$

as required.
(ii)

$\forall x \in \Omega \ \forall L_i \in LA \ \mu_{\neg L_i}(x) = \max \left(m_x(T) : L_i \notin T \right) =$
$\max \left(m_x(\emptyset), \max \left(m_x(\{L\}) : L \neq L_i \right), \max \left(m_x(T) : L_i \notin T, |T| > 1 \right) \right)$
$= \max \left(\epsilon, \max \left(\mu_L(x) : L \neq L_i \right), \max \left(f_\wedge \left(\mu_L(x) : L \in T \right) : L_i \notin T, |T| > 1 \right) \right)$
$= \max \left(\epsilon, \max \left(\mu_L(x) : L \neq L_i \right) \right)$

by the monotonicity of $f_\wedge$ as required. $\square$

8.3 An Axiomatic Approach to Generalised Appropriateness Measures

In this section we attempt to identify a set of axioms characterizing the notion of generalised appropriateness measure as outlined in the above sections of this chapter. To this aim we give an axiomatic definition of non-additive appropriateness degrees based on a weakening of the axioms in definition 55 (chapter 3).

DEFINITION 158 *(Axioms for Generalised Appropriateness Measures)*
A generalised appropriateness measure on $LE \times \Omega$ is a function $\mu : LE \times \Omega \to [0, 1]$ such that $\forall x \in \Omega, \ \theta \in LE \ \mu_\theta(x)$ quantifies the appropriateness of label expression θ as a description of value x and satisfies:

AM1 $\forall \theta \in LE$ *if* $\models \neg \theta$ *then* $\forall x \in \Omega \; \mu_\theta(x) = 0$

AM2 $\forall \theta, \; \varphi \in LE$ *if* $\theta \equiv \varphi$ *then* $\forall x \in \Omega \; \mu_\theta(x) = \mu_\varphi(x)$

AM3 $\forall \theta \in LE$ *there exists a function* $f_\theta : [0,1]^n \to [0,1]$ *such that*
$\forall x \in \Omega \; \mu_\theta(x) = f_\theta \left(\mu_{L_1}(x), \ldots, \mu_{L_n}(x) \right)$

AM4′ *For some t-conorm* $f_\vee$ *it holds that*
$\forall \theta, \; \varphi \in LE$, *if* $\models \neg \, (\theta \wedge \varphi)$ *then* $\forall x \in \Omega \; \mu_{\theta \vee \varphi}(x) = f_\vee(\mu_\theta(x), \mu_\varphi(x))$

Axioms AM1-AM3 are identical to those given in definition 55 (chapter 3). AM4′ is an extension of the additivity axiom AM4 in definition 55 to allow for non-additive disjunctive combination of incompatible expressions. The intuition behind this axiom is as follows: Since You need not take into account any logical dependencies between θ and φ then You can evaluate the appropriateness of $\theta \vee \varphi$ by application of a simple disjunctive function (t-conorm) to the appropriateness measures of θ and φ respectively.

We now show that there is a one to one correspondence between those measures satisfying axioms AM1-AM4′ as given in definition 158 and those define using generalised mass assignment as outlined in the first section of this chapter.

THEOREM 159 *(Characterization Theorem)*
μ *is a generalised appropriateness measure iff* $\forall x \in \Omega$ *there exists a generalised mass assignment* $m_x : 2^{LA} \to [0,1]$ *such that*

$$m_x = \Delta \left(\mu_{L_1}(x), \ldots, \mu_{L_n}(x) \right)$$

for some generalised mass selection function Δ *and*

$$\forall \theta \in LE \; \mu_\theta(x) = f_\vee \left(m_x(S) : S \in \lambda(\theta) \right)$$

Proof
($\Leftarrow$)
By theorem 33 (chapter 3) if $\models \neg \theta$ *then* $\mu_\theta(x) = f_\vee \left(m_x(T) : T \in \emptyset \right) = 0$ *and hence AM1 holds.*
$\forall \theta, \varphi \in LE : \; \theta \equiv \varphi$ *we have by corollary 32 that*

$$\forall x \in \Omega \; \mu_\theta(x) = f_\vee \left(m_x(T) : T \in \lambda(\theta) \right) =$$
$$f_\vee \left(m_x(T) : T \in \lambda(\varphi) \right) = \mu_\varphi(x)$$

Hence, AM2 holds.

$$\forall \theta \in LE \; \mu_\theta(x) = f_\theta \left(\mu_{L_1}(x), \ldots, \mu_{L_n}(x) \right) \; where$$
$$f_\theta \left(\mu_{L_1}(x), \ldots, \mu_{L_n}(x) \right) = f_\vee \left(m_x(T) : T \in \lambda(\theta) \right)$$

Hence, AM3 holds

If $\models \neg\,(\theta \wedge \varphi)$ *then by definition 25 and lemma 58 (chapter 3)* $\lambda(\theta \wedge \varphi) = \emptyset \Rightarrow$ $\lambda(\theta) \cap \lambda(\varphi) = \emptyset.$ *Hence by the associativity and commutativity of* $f_\vee,$ $\forall x \in \Omega$

$$\mu_{\theta\vee\varphi}(x) = f_\vee \left(m_x(T) : T \in \lambda(\theta) \cup \lambda(\varphi) \right) =$$
$$f_\vee \left(f_\vee \left(m_x(T) : T \in \lambda(\theta) \right), f_\vee \left(m_x(T) : T \in \lambda(\varphi) \right) \right) = f_\vee \left(\mu_\theta(x), \mu_\varphi(x) \right)$$

Hence, AM4' holds.

$(\Rightarrow)$

By the disjunctive normal form theorem for propositional logic it follows that

$$\forall \theta \in LE \ \theta \equiv \bigvee_{\alpha \in ATT_\theta} \alpha$$

Hence by AM2 and AM4' we have that

$$\forall x \in \Omega \ \mu_\theta(x) = \mu_{\left(\bigvee_{\alpha \in ATT_\theta} \alpha \right)}(x) = f_\vee \left(\mu_\alpha(x) : \alpha \in ATT_\theta \right)$$

Now $\forall x \in \Omega,$ $\forall T \subseteq LA$ *let*

$$m_x(T) = \mu_{\alpha_T}(x)$$

Now clearly m_x *is a generalised mass assignment. Also by lemma 58 (chapter 3)*

$$\forall \theta \in LE \ \mu_\theta(x) = f_\vee \left(\mu_\alpha(x) : \alpha \in ATT_\theta \right) =$$
$$f_\vee \left(\mu_{\alpha_T}(x) : T \in \lambda(\theta) \right) = f_\vee \left(m_x(T) : T \in \lambda(\theta) \right)$$

Finally, from AM3 we see that m_x *can be determine uniquely from* $\mu_{L_1}(x), \ldots, \mu_{L_n}(x)$ *according to the generalised mass selection function*

$$m_x(T) = \Delta \left(\mu_{L_1}(x), \ldots, \mu_{L_n}(x) \right)(T) = f_{\alpha_T} \left(\mu_{L_1}(x), \ldots, \mu_{L_n}(x) \right)$$

as required.

Theorem 159 means that for a fixed value of x a generalised appropriateness measure is characterised by a $f_\vee$-decomposable fuzzy measure (see [19] and [105]) M_x on $2^{2^{LA}}$ where

$$\forall A \subseteq 2^{LA} \ M_x(A) = f_\vee \left(m_x(T) : T \in A \right) \text{ and } \forall \theta \in LE \ \mu_\theta(x) = M_x \left(\lambda(\theta) \right)$$

8.4   The Law of Excluded Middle

As noted earlier generalised appropriateness measures are not guaranteed to satisfy the law of excluded middle, for which an additional normalization assumption is required. To understand this let us first consider the meaning of tautologies such as $\theta \vee \neg\theta$ in label semantics. In this context tautologies simply express Your belief that some subset of LA (including the empty set) is the set of appropriate labels for an instance. In other words, either the instance can be described in terms of the labels of LA or no labels are appropriate to describe it (i.e. x is undescribable). Hence, $\lambda(\theta \vee \neg\theta) = 2^{2^{LA}}$ and therefore

$$\forall x \in \Omega \forall \theta \in LE \ \mu_{\theta\vee\neg\theta}(x) = f_\vee \left(m_x(T) : T \subseteq LA\right)$$

From this we can see that $\mu_{\theta\vee\neg\theta}(x) < 1$ implies that You allocate a non-zero belief to the possibility that there are some labels outside of the current language LA that are appropriate to describe x. This is in effect an open world assumption regarding label definitions. Alternatively, if for every instance x $\mu_{\theta\vee\neg\theta}(x) = 1$ then You hold the closed world assumption view that if x is describable then it is describable in terms of the labels in LA.

Now let us consider the law of excluded middle with respect to additive and possibilistic appropriateness measures. For additive appropriateness measures the requirement that mass assignment values sum to one ensures that excluded middle holds. It should be noted, however, that for the more general class of appropriateness measure based on the t-conorm $f_\vee(a,b) = \min(1, a+b)$, where the normalisation condition for mass assignments is not insisted upon, the law of excluded middle is not guaranteed.

In general, possibilistic appropriateness measures do not satisfy the law of excluded middle except when certain conditions are satisfied. For instance, if for a given instance x $\max\left(\mu_L(x) : L \in LA\right) = 1$ then clearly $\mu_{\theta\vee\neg\theta}(x) = \max\left(m_x(T) : T \subseteq LA\right) = 1$. Hence, if the labels in LA cover the universe Ω such that for every x $\max\left(\mu_L(x) : L \in LA\right) = 1$ then excluded middle holds. Alternatively, in the case of possibilistic appropriateness measures based on the conjunctive possibilistic msf, we can guarantee excluded middle by ensuring that $m_x(\emptyset) = \epsilon = 1$ whenever $\max\left(\mu_L(x) : L \in LA\right) < 1$ (see theorem 157).

Summary

In this chapter we have investigated non-additive appropriateness measures where the mass assignment values on labels sets are aggregated disjunctively using a t-conorm. Such measures are in fact generalisations of the additive measures studied throughout this volume since the latter can be obtained by selecting the t-conorm $f_\vee(a,b) = \min(1, a+b)$ and insisting that mass assignment values sum to one. A number of properties of generalised appropriateness measures have been studied, in particular for the special cases where $f_\vee$ is an Archimedean or Strict Archimedean t-conorm or where $f_\vee = \max$. In the latter case, referred to as possibilistic appropriateness measures, we identify a family of generalised mass selection functions for which the appropriateness measures of a conjunction of labels corresponds to a t-norm applied to the appropriateness measures of the conjuncts. Finally, we give an axiomatic characterization of generalised appropriateness measures and then discuss how best to interpret the failure of the law of excluded for these non-additive measures.

References

[1] J. Aczel, (1966), *Lectures on functional Equations and their applications*, Academic Press, New York

[2] J. Aguilar-Martin, (2002), 'Independence and Conditioning in a Connectivistic Fuzzy Logic Framework' in *Soft Methods in Probability, Statistics and Data Analysis* (eds. P. Grzegorzewski, O. Hryniewics, M. A. Gil), Advances in Soft Computing, Springer Verlag

[3] T. Alsinet, L. Godo, (2000), 'A Complete Calculus for Possibilistic Logic Programming with Fuzzy Propositional Variable' *Proceedings of Uncertainty in AI 2000* (2000)

[4] J.F Baldwin, T.P. Martin, B.W. Pilsworth, (1995), *Fril - Fuzzy and Evidential Reasoning in A.I.*
Wiley, New York

[5] J.F. Baldwin, J. Lawry, T.P. Martin, (1996), 'A Mass Assignment Theory of the Probability of Fuzzy Events', *Fuzzy Sets and Systems* Vol. 83, No. 3 pp353-368

[6] J.F. Baldwin, J. Lawry, T.P. Martin, (1996), 'A Note on Probability/Possibility Consistency for Fuzzy Events', *Proceedings of the Sixth International Conference on Information Processing and Management of Uncertainty in Knowledge-Based Systems (IPMU 96)*, Vol. 1, pp521-526

[7] J.F. Baldwin, J. Lawry, T.P. Martin, (1998), 'A Note on the Conditional Probability of Subsets of a Continuous Domain', *Fuzzy Sets and Systems*, Vol. 96, No. 2, pp211-222

[8] R. Bellman, *Adaptive Control Processes: A Guided Tour*, Princeton University Press (1961)

[9] M. Black, (1937), 'Vagueness: An Excercise in Logical Analysis', *Philosophy of Science*, Vol. 4, pp427-455

[10] C.L. Blake, C.J. Merz, (1998), *UCI Repository of Machine Learning Databases*, http://www.ics.uci.edu/~mlearning/MLRepository.html

[11] D.I. Blockley, (1980), *The Nature of Structural Design and Safety*, Ellis Horwood, Chichester

[12] N. Bonini, D. Osherson, R. Viale, T. Williamson, (1999), 'On the Psychology of Vague Predicate', *Mind and Language*, Vol. 14, No. 4, pp377-393

[13] M.K. Breteler, A. Bezuijen, (1998), 'Design Criteria for Placed Block Revetments', *Dikes and Revetments: Design, Maintenance and Safety Assessment*, (Ed. K. W. Pilarczyk), pp217-48

[14] C.B. Brown, (1980), 'A Fuzzy Safety Measure', *ASCE J. Eng. Mech. Div.*, Vol. 105(EM5), pp855-872

[15] N.W. Campbell, B.T. Thomas, T. Troscianko, (1997), 'Automatic Segmentation and Classification of Outdoor Images using Neural Networks', *International Journal of Neural Systems*, Vol.8 No.1, pp137-144

[16] C. Cremona, Y. Gao, (1997), 'The Possibilistic Reliability Theory: Theoretic Aspects and Applications', *Structural Safety*, Vol. 19, No. 2, pp173-201

[17] W. Cui, D.I. Blockley, (1991), 'On the Bounds for Structural System Reliability', *Structural Safety*, Vol. 9, pp247-259

[18] D.C. Dennett, (1991), *Consciousness Explained*, Little, Brown and Company

[19] D. Dubois, H. Prade, (1982), 'A class of fuzzy measures based on triangular norms', *International Journal of General Systems*, Vol. 8 No. 1, pp. 43-61

[20] D. Dubois, H. Prade, (1988), 'An Introduction to Possibility and Fuzzy Logics' in *Non-Standard Logics for Automated Reasoning* (eds. P. Smets et al), Academic Press, pp742-755

[21] D.Dubois, H. Prade, (1990), 'Measuring Properties of Fuzzy Sets: A General Technique and its use in Fuzzy Query Evaluation', *Fuzzy Sets and Systems*, Vol. 38, pp137-152

[22] D. Dubois, H. Prade, (1994) 'Can we Enforce Full Compositionality in Uncertainty Calculi?', Proceedings of the 12'th National Conference on Artificial Intelligence (AAAI 94), Seattle, pp149-154

[23] D. Dubois, H. Prade, P. Smets, (1994), 'Partial Truth is not Uncertainty: Fuzzy Logic versus Possibility Theory', *IEEE Expert*, Vol.9, No.4, pp15-19

[24] D. Dubois, H. Prade, (1994), 'Fuzzy Sets - a Convenient Fiction for Modelling Vagueness and Possibility', *IEEE Transactions on Fuzzy Systems*, Vol. 2, pp16-21

[25] D. Dubois, H. Prade, (1997), 'The Three Semantics of Fuzzy Sets' *Fuzzy Sets and Systems* Vol. 90, pp141-150

[26] D. Dubois, H. Prade 'Possibility Theory: Qualitative and Quantitative Aspects' in *Handbook of Defeasible Reasoning and Uncertainty Management Systems* (eds. D.M. Gabbay, P. Smets) Vol. 1, Kluwer (1998) pp169-226

[27] D. Dubois, S. Moral, H. Prade, (1997), 'A Semantics for Possibility theory based on likelihoods' Journal of Mathematical Analysis and Applications Vol. 205, pp359-380

[28] D. Dubois, F. Esteva, L. Godo. H. Prade, (2001), 'An Information-Based Discussion of Vagueness', Proceedings of the 10'th IEEE International Conference of Fuzzy Systems

[29] I. Elishakoff, Y- Ben-Haim, (1990), *Convex Models of Uncertainty in Applied Mechanics*, Elesevier, Amsterdam

[30] C. Elkan, (1993), "The paradoxical Success of Fuzzy Logic" in *Proceedings of the Eleventh National Conference on Artificial Intelligence* MIT Press, pp698-703

[31] M.J. Frank, (1979) 'On the Simultaneous Associativity of $F(x,y)$ and $x + y - F(x,y)$', *Aequationes Math* Vol. 19

[32] B.R. Gaines, (1978), 'Fuzzy and Probability Uncertainty Logics' *Journal of Information and Control* Vol. 38, pp154-169

[33] J. Gebhardt, R. Kruse, (1993), 'The Context Model: An Integrating View of Vagueness and Uncertainty', *International Journal of Approximate Reasoning*, Vol.9, pp283-314

[34] R. Giles, (1979), 'A Formal System for Fuzzy Reasoning', *Fuzzy Sets and Systems*, Vol. 2, pp233-257

[35] R. Giles, (1988), 'The Concept of Grade of Membership', *Fuzzy Sets and Systems*, Vol. 25, pp297-323

[36] I.R. Goodman, (1982), 'Fuzzy Sets as Equivalence Classes of Random Sets' in *Fuzzy Set and Possibility Theory* (ed. R. Yager), pp327-342

[37] I.R. Goodman, (1984), 'Some New Results Concerning Random Sets and Fuzzy Sets', *Information Science*, Vol. 34, pp93-113

[38] I.R. Goodman, H.T. Nguyen ,(1985), *Uncertainty Models for Knowledge Based Systems* North Holland

[39] R.P. Gorman, T.J. Sejnowski, 'Analysis of Hidden Units in a Layered Network Trained to Classify Sonar Targets' *Neural Networks*, Vol.1, (1988), pp75-89

[40] S. R. Gunn, (2003), *Matlab support vector machine toolbox (version 2.1)*,   http://www.isis.ecs.soton.ac.uk/resources/svminfo/download.php

[41] P. Hajek, J. B. Paris, (1997), 'A Dialogue on Fuzzy Logic', *Soft Computing*, Vol.1, pp3-5

[42] P. Hajek, (1999), 'Ten Questions and One Problem on Fuzzy Logic', *Annals of Pure and Applied Logic*, Vol. 96, pp157-165

[43] J.W. Hall, J. Lawry, (2003), 'Fuzzy Label Methods for Constructing Imprecise Limit State Functions', *Structural Safety*, Vol. 28, pp317-341

[44] E. Hisdal, (1988) 'Are grades of membership probabilities' *Fuzzy Sets and Systems* Vol. 25, pp325-348

[45] E.B. Hunt, J. Marin, P.T. Stone, (1966), *Experiments in Induction*, Academic Press.

[46] W. Hurlimann, (2004), 'Fitting Bivariate Cumulative Returns with Copulas', *Computational Statistics and Data Analysis*, Vol. 45, pp355-372

[47] V.N. Huynh, Y. Nakamori, T.B. Ho, G. Resconi, (2004), 'A Context Model for Fuzzy Concept Analysis Based on Modal Logic', *Information Sciences*, Vol.160, pp111-129

[48] R. Hyndman and M. Akram, (2003) *Time series data library*, http://www-personal.buseco.monash.edu.au/~hyndman/TSDL/index.htm

[49] A.J. Izenman, (1985) 'J.R. Wolf and the Zurich sunspot relative numbers', *The Mathematical Intelligencer*, Vol.7 No.1, pp27-33

[50] R. C. Jeffrey, (1965), *The Logic of Decision*, Gordon and Breach, New York

[51] J. Kampe de Feriet, (1982), 'Interpretation of Membership Functions of Fuzzy Sets in Terms of Plausibility and Belief' in *Fuzzy Information and Decision Processes* (Eds. M.M. Gupta, E. Sanchez), North-Holland, pp93-97

[52] R.L. Keeney, D. Von Winterfeldt, (1991), 'Eliciting Probabilities from Experts in Complex Technical Problems', *IEEE Transactions on Engineering Management*, Vol. 38, pp191-201

[53] E.P. Klement, R. Mesiar, E. Pap, (2000), *Triangular Norms*, Vol. 8 of Trends in Logic, Kluwer Academic Publishers, Dordrecht

[54] G.J. Klir, B. Yuan, (1995), *Fuzzy Sets and Fuzzy Logic*, Prentice Hall

[55] I. Kononenko, (1991), 'Semi-Naive Bayesian Classifier', *Proceedings of EWSL-91, Sixth European Workshop on Learning*, Springer, pp206-219

[56] R. Kruse, C. Borgelt, D. Nauck, (1999), 'Fuzzy Data Analysis: Challenges and Perspectives', *Proceedings of the 8'th IEEE Conference on Fuzzy Systems (FUZZ-IEEE99)*

[57] A. Kyburg, (2000), 'When Vague Sentences Inform: A Model of Assertability', *Synthese*, Vol. 124, pp175-192

[58] M. Laviolette, J. W. Seaman, J. Douglas Barrett, W.H. Woodall, (1995), 'A Probabilistic and Statistical View of Fuzzy Methods', *Technometrics*, Vol. 37, No.3, pp249-261

[59] J. Lawry, (1998), 'A Voting Mechanism for Fuzzy Logic', *International Journal of Approximate Reasoning*, Vol. 19, pp315-333

[60] J. Lawry, (2001), 'Label Prototypes for Modelling with Words', *Proceedings of NAFIPS 2001 (North American Fuzzy Information Processing Society Conference)*

[61] J. Lawry, (2001), 'Query Evaluation using Linguistic Prototypes', *Proceedings of the 10'th IEEE Conference on Fuzzy Systems*, Melbourne, Australia

[62] J. Lawry, (2001), 'Label Semantics: A Formal Framework for Modelling with Words', *Proceedings of The European Conference on Symbolic and Quantitative Approaches to Reasoning Under Uncertainty, Lecture Notes in Artificial Intelligence*, (Eds. S. Benferhat, P. Besnard), Vol. 2143, pp374-385

[63] J. Lawry, (2002), 'Label Prototypes for Data Analysis', *Proceedings of the second UK Workshop on Computational Intelligence*, pp112-119

[64] J. Lawry, (2002), 'A New Calculus for Linguistic Prototypes in Data Analysis', *Proceedings of Soft Methods in Probability, Statistics and Data Analysis, Advances in Soft Computing*, (Eds. P. Grzegorzewski, O. Hryniewicz), Physica-Verlag, pp116-125

[65] J. Lawry, J. Recasens, (2003), 'A Random Set Model for Fuzzy Labels', *Proceedings of The European Conference on Symbolic and Quantitative Approaches to Reasoning Under Uncertainty, Lecture Notes in Artificial Intelligence* (Eds. T. Nielsen, N. Zhang), Vol.2711, pp357-369

[66] J. Lawry, (2003), 'Random Sets and Appropriateness Degrees for Modelling with Labels' *Modelling with Words* (Eds. J. Lawry, J. Shanahan, A. Ralescu), Lecture Notes in AI, Vol. 2873

[67] J. Lawry, (2004), 'A Framework for Linguistic Modelling' *Artificial Intelligence*, Vol. 155, pp1-39

[68] J. Lawry, J.W. Hall, R. Bovey, (2004), 'Fusion of Expert and Learnt Knowledge in a Framework of Fuzzy Labels', *International Journal of Approximate Reasoning*, Vol. 36, pp151-198

[69] D.D. Lewis, (1998), 'Naive Bayes at Forty: The Independence Assumption in Information Retrieval', Proceedings of ECML-98, 10th European Conference on Machine Learning, Lecture Notes in AI, pp4-15, Springer Verlag

[70] J.W. Lloyd *Foundations of Logic Programming* (Second Edition), Springer-Verlag (1987)

[71] W.P. Mackeown, P. Greenway, B.T. Thomas, W.A. Wright, (1994), 'Contextual Image Labelling with a Neural Network', *IEE Vision, Speech, and Signal Processing*, pp238-244

[72] H.T. Nguyen, (1978), 'On Random Sets and Belief Functions', *Journal of Mathematical Analysis and Applications*, Vol. 65, pp531-542

[73] H.T. Nguyen, (1984), 'On Modeling of linguistic Information using Random Sets' *Information Science* Vol. 34, pp265-274

[74] N.V. Noi, T.H. Cao, (2004), 'Annotated Linguistic Logic Programs for Soft Computing', *Proceedings of the 2nd International Conference of Vietnam and Francophone Informatics Research*,

[75] D.N. Osherson, E.E. Smith, (1981), 'On the Adequacy of Prototype Theory as a Theory of Concepts', *Cognition*, Vol. 9, pp35-58

[76] D.N. Osherson, E.E. Smith, (1982), 'Gradedness of Conceptual Combination', *Cognition*, Vol. 12, No. 3, pp299-318

[77] R. Parikh, (1996) 'Vague Predicates and Language Games', *Theoria (Spain)*, Vol. XI, No. 27, pp97-107

[78] J.B. Paris, (1994), *The Uncertain Reasoners Companion: A Mathematical Perspective*, Cambridge University Press

[79] J.B. Paris, (1997), 'A Semantics for Fuzzy Logic', Soft Computing, Vol. 1, pp143-147

[80] J. B. Paris, (2000), 'Semantics for Fuzzy Logic Supporting Truth Functionality', in *Discovering the World with Fuzzy Logic*, (Eds. V. Novak, I. Perlieva), Studies in Fuzziness and Soft Computing, Vol. 57, Physica Verlag, pp82-104

[81] J. Pearl, (1988), *Probabilistic Reasoning in Intelligent Systems: Networks of Plausible Inference*, Morgan-Kaufmann

[82] Z. Qin, J. Lawry, (2004), 'A Tree-Structured Classification Model Based on Label Semantics', *Proceedings of the 10'th International Conference on Information Processing and Management of Uncertainty in Knowledge-Based Systems (IPMU)*, pp261-268

[83] Z. Qin, J. Lawry, (2005), 'Decision Tree Learning with Fuzzy Labels', *Information Sciences*, Vol. 172, pp91-129

[84] J.R. Quinlan, (1986), 'Induction of Decision Trees', *Machine Learning*, Vol.1, pp81-106

[85] N.J. Randon, J. Lawry, (2002),'Linguistic Modelling Using a Semi-Naive Bayes Framework', *Proceedings of IPMU 2002 (Information Processing and Management of Uncertainty)*, France

[86] N.J. Randon, J. Lawry, (2003), 'Fuzzy Models for Prediction Based on Random Set Semantics' *Proceedings of EUSFLAT 2003 (European Society for Fuzzy Logic and Technology)*

[87] N. J. Randon, (2004), *Fuzzy and Random Set Based Induction Algorithms*, PhD Thesis, Bristol

[88] N. Randon, J. Lawry, (2004), 'Classification and Query Evaluation using Modelling with Words' to appear in *Information Sciences*

[89] E.H. Ruspini, (1991), 'On the Semantics of Fuzzy Logic', *International Journal of Approximate Reasoning*, Vol. 5, pp45-88

[90] B. Russell, (1923), 'Vagueness', *The Australasian Journal of Psychology and Philospsophy*, Vol. 1, pp84-92

[91] R.M. Sainsbury, (1995), 'Vagueness, Ignorance, and Margin for Error', *The British Journal for the Philosophy of Science*, Vol. 46, No. 4, pp589-601

[92] D.G. Schwartz, (1997), 'Dynamic Reasoning with Qualified Syllogisms', *Artificial Intelligence*, Vol. 93, No. 1-2, pp103-167

[93] D.G. Schwartz, (2000), 'Layman's Probability Theory: A Calculus for Reasoning with Linguistic Likelihood', *Information Sciences*, Vol. 126, No. 1-4, pp71-82

[94] B. Schweizer, A. Sklar, (1983), *Probabilistic Metric Spaces*, North-Holland, Amsterdam

[95] G. Shafer, (1976), *A Mathematical Theory of Evidence*, Princeton University Press

[96] J.G. Shanahan, (2000), *Soft Computing for Knowledge Discovery: Introducing Cartesian Granule Features*, Kluwer Academic Publishers, Boston/Dordrecht/London

[97] P Smets, (1990), 'Constructing the Pignistic Probability Function in a Context of Uncertainty', in *Uncertainty in Artificial Intelligence 5*, (Ed. M. Henrion), North Holland, Amsterdam, pp29-39

[98] R.A. Sorenson, (1988), *Blindspots*, Oxford, Clarendon Press

[99] S.F. Thomas, (1995), *Fuzziness and Probability*, ACG Press, Kansas

[100] F. Tonon, A. Bernadini, (1998), 'A Random Set Approach to Optimisation of Uncertain Structures', *Computers and Structures*, Vol. 68, pp283-600

[101] E. Trillas, (1979), 'Sobre Functiones de Negacion en la Teoria de Conjunctos Difusos', *Stochastica*, Vol.3, No.1, pp47-59

[102] P. Walley, (1991), *Statistical Inference from Imprecise Probabilities*, Chapman and Hall, London

[103] P. Walley, (1996), 'Measures of Uncertainty in Expert Systems', *Artificial Intelligence* , Vol. 83, pp1-58

[104] P. Walley, G. de Cooman, (2001), 'A Behavioural Model of Linguistic Uncertainty', *Information Sciences*, Vol. 134, pp1-37

[105] S. Weber, (1984), '⊥-Decomposable Measures and Integrals for Archimedean t-Conorms', *Journal of Mathematical Analysis and Applications* 101, pp114-138

[106] A.S. Weigend, B.A. Huberman, and D.E. Rumelhart, (1992), 'Predicting sunspots and ex-change rates with connectionist networks' *Non-linear*

Modelling and Forecasting (Eds. M. Casdagli and S. Eubank) , SFI Studies in the Sciences of Complexity Proceedings, volume XII, pages 395-432, Addison-Wesley

[107] T. Williamson, (1992), 'Vagueness and Ignorance', *Proceedings of the Aristotlian Society*, Vol. 66, pp145-162

[108] T. Williamson, (1994), *Vagueness*, Routledge

[109] I. H. Witten, E. Frank, (1999), *Data Mining: Practical Machine Learning Tools and Techniques with Java*, Morgan Kaufmann, available from http://www.cs.waikato.ac.nz/~ml/weka/

[110] L.A. Zadeh,(1965), 'Fuzzy Sets', *Information and Control*, Vol.8, No.3 pp338-353

[111] L.A. Zadeh, (1968), 'Probability Measures of Fuzzy Events' *Journal of Mathematical Analysis and Applications* Vol. 23, pp421-427

[112] L.A. Zadeh 'The Concept of Linguistic Variable and its Application to Approximate Reasoning Part 1' *Information Sciences* Vol. 8 (1975) pp199-249

[113] L.A. Zadeh 'The Concept of Linguistic Variable and its Application to Approximate Reasoning Part 2' *Information Sciences* Vol. 8 (1975) pp301-357

[114] L.A. Zadeh 'The Concept of Linguistic Variable and its Application to Approximate Reasoning Part 3' *Information Sciences* Vol. 9 (1976) pp43-80

[115] L.A. Zadeh 'Fuzzy Sets as a Basis for a Theory of Possibility' *Fuzzy Sets and Systems* Vol. 1 (1978) pp3-28

[116] L.A. Zadeh "A Note on Prototype Theory and Fuzzy Sets" *Cognition* Vol. 12 (1982) pp291-297

[117] L.A. Zadeh 'Fuzzy Logic = Computing with Words' *IEEE Transactions on Fuzzy Systems* Vol. 4 (1996) pp103-111

[118] L.A. Zadeh, (2004), 'Precisiated Natural Language (PNL)', *AI Magazine*, Vol. 25, No. 3, pp74-91

Index